Environmental Ethics For Engineers

Alastair S. Gunn

Department of Philosophy
University of Waikato
Hamilton, New Zealand

P. Aarne Vesilind

Department of Civil and
   Environmental Engineering
Duke University
Durham, North Carolina

 LEWIS PUBLISHERS, INC.

Library of Congress Cataloging-in-Publication Data

Gunn, Alastair S.
    Environmental ethics for engineers.

    Includes bibliographies and index.
    1. Environmental engineering—Moral and ethical
aspects. I. Vesilind, P. Aarne. II. Title.
TD153.G78   1986   174'.962   86-7235
ISBN 0-87371-074-6

LEWIS PUBLISHERS, INC.
121 South Main Street, P.O. Drawer 519,
Chelsea, Michigan 48118

PRINTED IN THE UNITED STATES OF AMERICA

Preface

WHEN THE STORY of the human race comes to be written, it will not be merely the story of *Homo sapiens*, rational man, for we are also *Homo faber*, handyman, the toolmaker, the technologist. Humans have achieved preeminence on earth because of the interacting capacities of our brains *and* our hands. One without the other would be inadequate even for physical survival.

The practice of engineering requires us to use both our brains and our hands. If science is the attempt to understand the world, and technology the attempt to fix what ails it, engineering encompasses both, and more. The discipline of engineering is not devoted to pure research, nor to fixing things, but to designing and building structures, systems and processes to sustain and enrich our lives.

This book is written for engineers who, while appreciating the contributions their profession has made to human well-being, also sense that the practice of engineering can raise some disturbing questions. For example:

- Do engineers overemphasize the technological function, ignoring or forgetting the duty to sustain and enrich our lives?
- Are engineers merely hired hands in the service of corporations or other sectional interests, working as mercenaries without worrying about the end they are meant to serve?
- Has human well-being been narrowly conceived in materialistic terms, ignoring other dimensions of human life such as beauty, truth and justice?
- Do engineers regard the service to *humans* as the only form of ethical behavior, with no thought for the rest of nature except as a resource?

Engineers harness technology to achieve their goals, yet the history of recent technology is a history of the wasteful use of mostly non-renewable resources, aimed at satisfying short-term human wants without regard either to future human needs or the rest of nature.

Although it is tempting to blame engineers for this short-sightedness, it is both inaccurate and unseemly to make engineers the scapegoats for our present environmental problems. Engineers have responded to the demands of society by developing energy sources, transportation and communication systems, new chemicals and medical skills, and so on, because that is what society has trained them to do. Like all professions and trades, engineering provides a service for which there is a demand.

Yet the profession is not totally blameless, since engineers *do* shape the society in which they function. The engineering profession is, potentially, more pervasive and interdisciplinary than any of the other professions.

Included in the menu of skills engineers must recognize, if not possess, is the ability to deal with situations where values conflict. For example, engineers may be required to choose between the value of an unspoiled natural vista which provides enjoyment and pleasure for sightseers; and the value of a new bridge which allows greater access to a remote valley but which visually intrudes and spoils the beautiful view. Or, engineers may be placed in the position of having to evaluate the relative merits (values) of using animals for the testing of cosmetics; versus providing untested chemicals for human use. Or, engineers may be required to design advanced weapons systems; which they might personally deplore. In every branch of engineering, conflicts arise which require moral judgment. Where these conflicts impact the natural environment, ethics related to environmental concerns become important. It is in the hope of stimulating greater environmental awareness within the engineering profession that we offer an environmental ethics book for engineers.

This book is divided into two parts. The first part is a primer on professional ethics as applied to the environment. Much of this is factual information, reviewing the organization of ethical thought applied to environmental concerns and tracing the roots of this ethical thinking. The connection between engineers and how they might interact with nature is made and it is shown that engineers must learn to understand these conflicts and apply the principles of ethics to engineering practice. We readily admit to selective use of what we consider important in the discussion of environmental ethics and make no pretense of presenting this review as being either comprehensive or balanced. The purpose of the first section of this book is merely to enhance the knowledge of professional and environmental ethics so that this information may be used to obtain a fuller understanding of the various articles which comprise the second section. Some of the nine chapters in the second part are significant in the development of environmental ethics, while others deal with controversial issues and professional approaches to ethics.

We have used this book, in manuscript form, as supplemental reading in our environmental engineering classes at Duke University. The discussion of ethics is usually reserved for the final few days of class, when the students should start asking "so what?" about the course material. We respond to this question by covering the principles of ethics in one lecture and spending two or more sessions discussing various readings. Engineering students who have spent four years learning how to crunch numbers and to solve technical problems to three significant figures admit that the study of environmental ethics introduces new and exciting concepts into their professional thinking, and provides a perspective which otherwise would be missing from their education.

The preparation of this book was facilitated by a study leave granted to the senior author from the University of Waikato, a leave spent at Duke University. The time for final editing and re-editing by the junior author was made possible by a sabbatical leave granted by Duke University. Much of this work was performed in the dungeons of the University College London, and grateful appreciation is extended to Professor Ken Ives of the Department of Civil and Municipal Engineering.

Friends who read various editions of the manuscript and of-

4

fered valuable advice include John Fielder of Villanova University, Tom McCollough of Duke University, and Elizabeth Endy of Cabrini College.

A.S. Gunn
P.A. Vesilind

Durham, N.C.
January 1986

Table of Contents

PART 1   FUNDAMENTALS

1.1   Ethics and Ethical Conduct in Engineering Practice   7
1.2   Environmental Ethics   17
1.3   Environmental Ethics and Professional Engineering   33

PART 2   READINGS AND CASE STUDIES

2.1   The Land Ethic   *Aldo Leopold*   43
2.2   The Tragedy of the Commons   *Garrett Hardin*   55
2.3   The Kepone Tragedy   *W. Goldfarb*   63
2.4   The Hooker Memos   *CBS News*   73
2.5   The Bunker Hill Lead Smelter   *C. Tate*   87
2.6   The Existential Pleasures of Engineering
        Samuel Florman   91
2.7   Decision Making in the Corps of Engineers
        P. A. Vesilind   101
2.8   Moral Development and Professional Engineering
        Elizabeth M. Endy and P. A. Vesilind   111
2.9   Should Trees Have Standing?   *Christopher D. Stone*   123

APPENDIX
A.1   Code of Ethics   *American Society of Civil Engineers*   131
A.2   Primer on Ethical Theories   137

Index   151
Author Biographies   157

To Our

Estonian
Latvian and
Lithuanian

ancestors

Part I
Fundamentals

1.1 Ethics and Ethical Conduct in Engineering Practice

"**I**T'S SO complicated!"

In the movie *The China Syndrome* a television reporter (Jane Fonda) is about to interview a nuclear power plant engineer (Jack Lemmon) who has decided to blow the whistle on what he believes to be an unsafe reactor, when his employer has the SWAT team shoot him down. Just before the shooting, Lemmon shakes his head in dismay and frustration, and says: "It's so complicated!"

He does not mean only that the nuclear reactor and the technical problems associated with the plant are complicated. He also means that his very act of speaking directly to the press, much against his employer's wishes, places him in a complex ethical quandary, a situation which he has neither anticipated nor is prepared to cope with. On the one hand, he knows that speaking out about the unsafe reactor is in the best interest of society, but on the other hand, accepted professional engineering practice dictates loyalty to the employer. He thus cannot do both — speak out and keep quiet — and thus finds himself in a situation where values conflict. What should he do if he is to act ethically?

Although society expects engineers to act ethically, difficulties arise in defining what is meant by *anyone* being "ethical," and what indeed is "ethics."

what is meant by "ethics"?

What is meant when a person is said to be ethical? There seem to be three possibilities:

- An ethical person is one who has any set of values and lives by them;
- An ethical person is one who has any set of values which are also shared by a group;
- An ethical person is one who lives by a set of values which are universally valid.

1. In the most general and least evaluative sense, an ethic is any set of values. An ethical person is one who makes decisions on the basis of some such set, and to act ethically is to behave consistently with one's professed values. Almost any set of values not grotesquely self-contradictory or otherwise incoherent is thus an ethic, and everyone except small children, psychopaths and hypocrites is ethical.

If there is no more to an ethic than self-prescription, ethical disagreement is impossible. For example, some of us may believe that bludgeoning baby seals is wrong, while others think it is acceptable, but neither party has any basis upon which to condemn the other.

No doubt it is good that everyone should hold some values and live by them. If individual responsibility and autonomy are good, then the development of personal values is good. Yet it doesn't matter very much to us whether others are ethical in this sense alone, except that life is easier if other people are predictable (which in this sense of "ethical" they will be). To advocate ethical behavior is not merely to wish that everyone adopt a personal code of ethics, regardless of what it is. Living by one's values seems to be a minimum, necessary and purely formal requirement of ethical behavior, not a sufficient, substantial requirement.

2. An ethic might be viewed as any set of values shared by a group, such as a community, church, society or profession. An ethical person is one who in practice adheres to such a set, and to act ethically is to act according to the values of the group in question.

Everyone is a member of society. Engineers, for example, benefit in various ways from membership in a range of groups — local community, school, profession, state, nation and so on. It is only fair that in recognition of the benefits thus received the engineers should accept burdens and obligations. These include obeying the rules of the community (e.g. criminal law), the school (e.g. the honor code), or the profession (e.g. the *Code of Ethics*). This is not because the rules are perfect or even especially good, but because as a citizen, student or professional, one has an obligation to obey the appropriate rules. Obviously, this view is most plausible in an egalitarian and democratic society, where the rules on balance benefit everyone and are justly administered, and where everyone has a say in the making of rules. It is least plausible (if at all) in an inegalitarian dictatorship, where the rules are constructed by those in power in order to benefit themselves at everyone else's expense.

Persons who do not accept a general duty to obey the rules in a just society are unethical because they are benefiting unfairly at the expense of those who do obey the rules. For example, a contractor in the United States who bribes a public official to obtain orders or to have safety violations overlooked is unfairly benefiting at the expense of honest competitors, while the corrupt official benefits unfairly at the expense of honest officials. In the event that a hazard to the public results, then the parties to the bribe have benefited unfairly by foisting social costs onto an unaware public.

This view of ethics has some advantages over the first concept of ethics. The acceptance of a shared set of values presumably tends to benefit everyone by reducing tensions and conflicts and encouraging trust and security. Shared values will often reflect discussion, reason and compromise.

It might appear that ethical disagreement, the notion of which could not arise in the first view of ethics, is possible with this concept. So it is, though is is limited to questions of what the code says or means, and how it is to be interpreted and applied. Hence it is not possible to make comparisons *between* codes, nor indeed to criticize ethically any clear provisions of the code.

Consider this example: a small country is ruled by a corrupt (by western standards) oligarch. All opposition, organized or otherwise, is illegal. The secret police are entrusted with the arrest, interrogation, and prosecution of suspects. Suspects are frequently beaten and tortured under interrogation or mysteriously disappear. The chief of the secret police defends these practices on the grounds of necessity: the alternative is chaos and anarchy. She also explains that her officers act on very strict rules. For example, no one under the age of 16, or over the age of 60 is to beaten or tortured. No one is to be arrested or liquidated unless there is some evidence that they have committed a serious crime (e.g., harboring a grudge against the regime). In short, there exists a strict professional code of practice which all police officers are expected to follow.

Also in this country, it is the custom for engineering and other contracts to be awarded partly on the basis of bribes. Bribing is not, of course, a sufficient condition of acceptance since public officials such as building inspectors are held personally liable for damage or loss of life due to violations of building codes. But a contract bid will not even be opened until several officials have been bribed; and the successful contractor will find it impossible to meet completion dates (and avoid penalties) without constant bribes to labor union leaders, customs officials and so on.

This is not precisely a code of practice. It is, nonetheless, the way things are done, and there is a rough consistency to the system. An experienced contractor knows just whom to bribe, how often, and how much. Anyone who fails to act according to the system fails to get business. Anyone who tries to blow the whistle will be delivered to the secret police.

Our condemnation of these systems is not limited to the claim that they would be unethical in the United States. We might want to say that the behavior of the police chief is partly excusable, in that she sincerely believes that this is the only way to proceed, whereas we would not in any way excuse such behavior on the part of the FBI, for example. But we also want to say that her behavior is *wrong*, that this society (judged by these two examples) is a bad society. Indeed, we want to say that Nazi concentration camps, Russian "mental hospitals" for dissidents and the South African apartheid system are wrong *regardless* of what Nazis, Commissars and the Afrikaaners think. But if to be ethical means *merely* to follow one's society's code, we are debarred from such judgment.

To advocate ethical behavior cannot then be simply to wish that everyone adopt some group code of ethics, regardless of what it is. Again living by some shared set of values is a formal, not a substantive condition of being ethical. Indeed, it may not even be that. An ethical pioneer such as Socrates or Thoreau might consider himself to be the only ethical person in his society.

3. The third view of ethics introduces a set of values to a system of conduct. It is true that there may not be any universally held values, but are there not some values which transcend society, which we can accept as criteria for evaluating both personal and social codes of ethics? It might be said that there are certain values which are better than others, values which we ought to accept even if these are rejected by the society. To act ethically, to be an ethical person, would then be to act according to these values.

Reformers, advocates of civil disobedience, and whistle blowers believe that, in certain cases, their values are superior to those prescribed by the code or rules under which they are expected to live and practice. They are not necessarily claiming that their values are universally valid or absolute — only that they are better than the accepted ones. The Western democratic tradition puts great value on justice, fairness, equality, democracy, autonomy and responsibility. We believe that these are good values, and that societies (including our own) should be evaluated according to the extent they promote such values.

If there are values which transcend society, we can appeal to them in order to criticize some societies (including our own) and to make suggestions for improvements. Ethical agreement and disagreement now depend on our answers to questions about the nature, interpretation, and application of such values. The result, in some cases at least, will be a judgment that a widespread practice is wrong, because it is inconsistent with a basic value. For example, the bludgeoning of baby seals is wrong because it violates the rights of the seals not to be made to suffer unnecessarily (and items made of baby seal are not "necessary"). Torture of suspects is wrong because it violates fundamental rights such as due process and equity before the law. The bribing of public officials is wrong because it is unfair and violates the trust placed in public officials.

In a very unjust system, a failure to conform to the code is thus not unethical. Very unjust societies are typically neither egalitarian nor democratic, and if a person gets no benefits (but rather suffers) from the rules, and has no say in their making, it is difficult to see how she or he could have an obligation to obey them. In the case of an unjust dictatorship, for example, we would *admire* a person who courageously tries to organize opposition to the regime, helps smuggle wanted political activists out of the country, and so on. We might well condone acts which, in our own country, we would condemn, such as theft, sabotage, or even assassination. An ethical person in such a society, it seems, would be one who obeys the law only when he or she has to, or when (by coincidence) it seems to be a fair law. But in such a society, citizens would have no ethical obligation to obey the law as such.

Similarly, engineers need not slavishly conform to accepted codes and laws. The very premise upon which the duty of obedience is based is that the rules are fair and reasonable. Where they are not, any citizen has a right, or more likely a *duty*, to try to get unjust or unreasonable rules changed. In extreme cases, where the existing rules are permitting or perpetuating a great wrong, citizens may be justified in breaking a rule, by civil disobedience or whistle blowing, for example.

The application of these values in the engineering profession is, however, fraught with difficulty, the most glaring of which is the almost total ignorance of ethical theory.

why teach ethics to engineers?

Ethical questions arise constantly in engineering professional practice. Engineers without a knowledge and background in this area confronted by an ethical problem have one of three options:

- Call in a consultant
- Ignore the problem
- Learn to deal with such situations themselves.

1. One obvious solution to any problem faced by engineers is

to call in a consultant. If you want to find out about population trends or fossils or data processing, you consult a demographer, a paleontologist or a computer scientist. So if you want to find out about values you consult an … an … ethicist?

The term "ethicist" implies that the solution to value-laden problems is reducible to a series of questions which, using logical and codified methods of analysis, certain highly trained people are equipped to answer. Just as the average person has no pretensions to understand very much about selective breeding or land drainage (so better call in a geneticist or a hydrologist), it is implied that ethics, too, can be understood only by specialists. And just as we expect our consultant geneticist or hydrologist to come up with the *right* answers, presumably an ethicist also has the truth for us.

Unfortunately, this approach will not work. Whether or not there are "right answers" in ethics, we argue above that some values are superior to others. Yet there are different, competing value systems, which need to be taken into account, and often even basic values such as equality and autonomy conflict. While in conventional fields of expertise there might be a standard state-of-the-art approach to solving problems, the ethicist can only give you one of several possible views, or present you with the choices. An ethicist, if indeed there is such a person, can not *solve* a problem.

2. A second approach is to ignore all value judgments and simply perform technical tasks which are requested by the client.

Consider, for example, hazardous waste management, where engineering problems such as the analysis, transportation, storing and reprocessing of hazardous waste materials require many engineering skills. If a secure landfill is to be designed, a consulting engineer might be hired to design a facility that will meet the client's requirements at lowest cost. This engineer could ignore all ethical questions by simply performing a technically proficient job, and staying within the law. For example, the engineer could specify, design and install liners for the landfill which meet all state and federal codes without informing the regulatory authorities that there are suspected geological formations at the site which might make these liners worthless and would create an eventual groundwater pollution problem. If the client is informed of this and chooses to ignore the warnings, the engineer (according to this approach to professional ethics) is no longer responsible for the consequences.

This form of engineering practice, where the law is followed and the client's best interests are protected without the complications of value-laden concerns, is promoted by some prominent engineers (see page 91). By this approach, ethical questions can be ignored as long as laws are not broken and the codes of professional conduct as published by the various engineering societies are adhered to.

But it should be perfectly clear that engineering problems are *not* purely technical challenges. Any engineering decision affects the welfare of someone other than the engineer and his/her employer or client — not to mention the rest of nature. For example, the design of more effective nozzles for spraying Agent Orange in Vietnam was an interesting and challenging engineering problem. If only technical considerations are important, the engineers who designed these nozzles should have been proud of their accomplishment, free of any concern as to the ultimate use of the nozzles.

In defense of the profession, such a lack of sensitivity in today's engineers is difficult to imagine. Engineers generally are quite cognizant of the secondary impacts of their decisions, and take them into account in the formulation of the solutions to problems. Thus ignoring ethics is not a valid approach for resolving value-laden professional problems.

3. A third alternative to resolving engineering problem where values conflict is to learn to cope with such situations and to make enlightened and well-considered decisions. This of course is the most difficult of the three options to implement, and yet seems to be the alternative of choice. As is the case with any other engineering tool such as calculus or fluid mechanics, ethical thought and decision-making skills should be a part of the engineer's repertoire. And as is the case with other engineering tools, ethics should be taught along with the other skills.

There is of course a "red herring" that pops up whenever it is suggested that ethics be taught. "Mind control!" cry the detractors. But there is a very significant difference between teaching *what* to think, and *how* to think. Teaching ethics in the classroom involves the expanding of the mind's horizons to encompass the wisdom of the ages, and to bring these *methods of thinking about difficult topics* to bear on contemporary problems. Teaching ethics is the exact opposite of mind control; it is the unshackling of the mind and making it free to understand other ideas and other values.

Ethical consideration, we believe, should be integrated into many existing engineering courses. Students in a course on hazardous waste management, for example, could be presented with questions about risk assessment, public health and safety, social justice in the distribution of benefits and burdens, obligations to future generations and responsibility for the rest of nature. Ethical considerations in hazardous waste management should not be offered as a separate, ancillary, and often optional topic, but as part of the study of techniques of secure landfill or the analysis and identification of wastes. These students, the future practicing professional engineers, would then become accustomed to considering ethical questions on the job. They would see public policy questions and technical questions as different aspects of the same problem, not as two separate problems requiring discrete answers.

We return to the problem of professional ethics, and the education of engineers in ethical thinking, in Chapter 1.3, following a discussion of a special kind of ethics; environmental ethics.

1.2 Environmental Ethics

ETHICAL QUESTIONS are questions about how we ought to act. We do not speculate, discuss and disagree about ethics merely for the fun of it, but because we wish to know how we should live, and particularly how we should treat others. But *which* others? If respect is a duty, who or what is worthy of respect? *Whose* happiness should we consider? In other words, what are the limits of the *moral community*, the class of beings which we ought to consider as having worth and towards which we are obligated to act or forbear in various ways? Most people today might answer that the human race is the entire moral community. In the discussion that follows, we question this view.

attitudes and exploitation

How we treat the rest of nature, and how we perceive it, are closely linked. In this section we explain this in a context of human relations, and move on to apply it to our dealings with nature.

Our values, and therefore our actions, are closely tied in with our perceptions. What in ordinary language is often called a person's *attitude* towards something may be described as a combination of how they perceive it and how they think it ought to be treated. The language we use is often an indication of our attitude, as well as a device for changing the attitudes of others.[1] People who refer to "niggers" or "honkies" indicate a perception of a racial group as inferior, and an intent or desire to discriminate. More subtly, to refer to duck or deer as "game" indicates a view of wildlife as a resource and approval of the killing of species thus designated.

To focus on the attitudinal or "motive" aspects of ethical language may devalue the role of reasoning in ethics. In some cases, the cynic might view the arguments advanced as rationalizations, conveniently invented to justify exploitation or the status quo. Examples are not hard to find.

A recent case of pseudo-scientific rationalization and selective use of data is the campaign by the electrical power industry to trivialize the problem of acid rain. There is overwhelming evidence that coal burning produces sulfur dioxide which is oxidized to sulfur trioxide and which in turn combines with water vapor to produce sulfuric acid.

Expressed in a greatly simplified way,

$$SO_2 + O \rightarrow SO_3 + H_2O \rightarrow H_2SO_4$$

Lakes and streams in Canada, Northern USA and Scandinavia have been severely impacted by this precipitation and are no longer able to support fish life. The only solution seems to be the removal of sulfur from power plant emissions, but this will cost the industry millions of dollars. They have fought this solution by lobbying for legislative relief and by initiating a public relations campaign. A major effort in this campaign is the publication of a booklet[2] in which it is argued:

a. Rain is today no more acid than it was years ago. But in case it *is* more acid, ...
b. Sulfur oxide emissions did not make it more acid. But in case they *did* make it more acid, ...
c. Power plants are really not responsible for the sulfur in the atmosphere. But in case they *are* responsible, ...
d. Sulfur and higher acidities in rain are actually *beneficial* to plants and amimals!

Examples of this sort will tempt us to believe that ethics is no more than the dressing up of prejudices and special interests with arguments to make them appear respectable. But our ethical beliefs and arguments should not be dismissed as mere bias and cynical devices to defend our self-interest. We may properly condemn attempts to defend what is wrong by devising theories and inventing facts to make one's actions appear right. But to have a clear conscience, and to defend our actions to others, we would like to believe that our actions are ethical. The desire to be ethical and to have the approval of others is surely not itself unethical.

A disparity between thought and action creates uncomfortable feelings of what psychologists call "cognitive dissonance." The desire to reduce dissonance is legitimate, provided it is exercised honestly. Moreover we want our behavior to be defensible as a rational response to how the world really is. We are most comfortable when we can justify our behavior by appealing to facts. Thus our belief in the equality of men and women can be defended by showing that men and women, given equal opportunities, make equally good engineers. The giving of reasons of this sort helps guard against prejudices, and a sincere respect for the truth prevents ethics from degenerating into mere self-serving rationalization.

In this spirit, let us consider the institution of slavery. If a person were socialized into a slave-owning society, it would be difficult to justify the owning of slaves and yet believe that one's slaves were even potentially one's equals in intellect or character. In sophisticated slave-owning societies, therefore, pseudo-scientific theories arise to "justify" the institution of slavery. Thus Aristotle tried to justify slavery on the premise that everyone is either a "natural" slave or master. He described slaves as typically of large stature and low intelligence, and masters as highly intelligent but unable to carry out heavy work. Aristotle believed that since the ability to reason and plan and the strength to carry out plans are not found in the same person, slavery is good for masters and slaves alike. He also believed that the master-slave distinction applied in barbarian (non-Greek) societies, but that, relative to Greeks *all* barbarians could properly be regarded as slaves.

A sincere believer in Aristotle's theory of slavery can feel no community with a slave, but instead can justify what we would see as exploitation by appealing to "facts;" slaves and masters need each other. Masters benefit most, but then they have more potential. A slave is merely a useful living tool, like an ox but more intelligent.

Aristotle's beliefs about slavery are ideological, in the sense that they appear to function as an apology for, or rationalization of, an oppressive, unjust institution. But we should not rush to accuse him of insincerity. Rather, we should apply the insights gained from studying Aristotle to our own ethical beliefs. Sexists, for instance, have traditionally viewed men as decisive, rational, calm, natural leaders and protectors; women as dithery, irrational, emotional beings in need of care and protection. Like Aristotle's theory of slavery, then, sexism illustrates the fact that exploitation is generally accompanied by or tied to theories explaining the supposed inferiority of the exploited group.

In extremely inegalitarian societies, ethical questions concerning the relation of superior to inferior never arise. The exploiting class does not regard the exploited class as part of the moral community. In very sexist or racist societies, women or minorities are viewed as mere chattels, to whom no obligations are owed.

Moral progress may be said to begin when members of the dominant class start to see themselves as having duties toward their inferiors. Men and women, black and white, cannot form a moral community unless they are viewed as equals, with a right to mutual respect. At most, racists view minorities as moral *patients* (they are the *object* of obligations) but not as *agents* (they do not *have* obligations). In contrast, a moral community is a society of equals, characterized by mutual respect.

A parallel question about the scope of the moral community concerns human attitudes toward the rest of nature. Until comparatively recently, almost everyone in the West thought of *ethics as concerned exclusively with relations between humans.* In the last century or so it has gradually been recognized that we have some obligations towards the non-human world. To cause gratuitous suffering to animals, for example, is generally thought to be wrong. This ethic is very selective, however, and does not extend to the "lower" animals, let alone plants. Our obligations concerning the rest of nature, such as they are, are usually seen as part of our duty to humans. For example, we recognize our duty to preserve wilderness areas for aesthetic enjoyment and scientific study, and to conserve resources for future generations. We might call this "enlightened" view the *garden ethic.*

At best, we have an *ethic about the environment,* much as enlightened racists believe themselves to have obligations towards minorities. The modern equivalent of the nineteenth century British imperialists' "white man's burden" might be called the "human burden." The recently popularized terms "human chauvinist" (like "male chauvinist") and the unfortunate tongue twister "speciesist" (like "racist" and "sexist") indicate this parallel. We do not see ourselves as part of a moral community with the rest of nature. Indeed, we usually distinguish sharply between "mankind" and "nature." What is the origin of this attitude?

historical roots: religious

One source of the dichotomy between humanity and nature, it is argued, is religious. Lynn White, Jr.,[3] suggests that the Judeo-Christian tradition is *the* source of our present attitude towards nature, and therefore of our exploitative treatment of the environment. The dominant tradition is found in Genesis I:28. "God blessed them, and said to them, be fruitful and increase, fill the earth and subdue it, rule over the fish in the sea, the birds of heaven, and every living thing that moves upon the earth." This seems to be fairly explicit about the relative roles of humans and the rest of creation. White believes that "especially in its Western form, Christianity is the most anthropocentric religion the world has seen," and that Christianity inherited from Judaism the view that "God planned (the creation) explicitly for man's benefit and rule; no item in the physical creation had any purpose save to serve man's purpose."

In contrast, some Christian writings are what we may call "benignly speciesist." Thus Bishop Hugh Montefiore[4] finds in the Old Testament a view that God intends humans to be "stewards and trustees for God," so that humans have "an inalienable duty towards and concern for their total environment, present and future; and this duty towards environment does not merely include their fellow-men, but all nature and all life."

historical roots: secular

Whether or not White is correct, we must also consider a secular source of our attitude toward nature. Since at least the seventeenth century, the dominant secular stream of Western culture has sharply separated the human from the non-human by ascribing unique properties to humans. The *mechanistic* view of nature is associated with such philosophers as Galileo (1564-1642), Francis Bacon (1561-1621), Thomas Hobbes (1588-1679), and, in some interpretations, René Descartes (1596-1650). The world, and all that dwells therein, is viewed as organized material and complex machines. Natural processes — including the behavior of animals — are explicable causally. The whole of nature is a deterministic system. The human body is a machine, too: blood flows, food is digested, the senses receive impressions.

Some mechanists, such as Hobbes, regarded humans as no more than machines. Others see humans as unique, because they have the ability to reason. The Cartesian tradition, for example, credits humans with a non-material soul as well as a body, unlike other living things which are merely bodies. Some Cartesians concluded that only humans can think, have emotions, make choices, or feel pleasure and pain. Since animals are mere machines, this argument continues, we cannot harm them except in the sense that one might damage a clock. Non-humans do not have interests and therefore do not deserve respect. Our treatment of animals, like our treatment of inanimate objects, is of no ethical significance. Seventeenth century scientists carried out excruciatingly painful experiments on animals, secure in the knowledge that the animals did not suffer. (Their cries of pain were regarded as mechanically produced, like the squeak of an unlubricated wheel.) An echo of this mechanistic view of animals is seen in the belief of many fishermen that fish are unable to feel pain.

modern speciesism

Most people today do not believe that animals are insensitive to pain, but they do believe that animals lack other "mental" characteristics. Many people condemn bullfighting, for example, but see nothing wrong with intensively raising animals for food, so long as no pain is caused. In fact intensive raising causes intense frustration, discomfort, deprivation and boredom, but this is not recognized by most people because they do not perceive pigs, calves, or chickens as being intelligent, curious or social.

Ignorance about animal behavior is paralleled by ignorance about what happens to animals. Most people never see what goes on in intensive rearing facilities, in slaughterhouses, and in the transport of livestock. They perceive the animal industry as humane, and therefore feel comfortable about eating meat. Likewise, they do not realize that new products (even non-useful ones such as cosmetics) are routinely tested on animals; that millions of animals suffer and die every year in such testing. They therefore feel comfortable going out and buying new products. Medical research, finally, is seen as providing great benefits for humans at the cost of minimum, unavoidable suffering to animals.

Most of us tend to believe that our use of animals does not cause suffering; that thanks to anti-cruelty and humane slaughter legislation animal suffering has largely been abolished. We do not recognize any animal interests other than avoiding physical pain. Just as Aristotle believed that slavery did not harm slaves, we believe that in general we do not harm animals. Where we *do* evidently harm them, we accept this because we believe that important human interests — life and health — are at stake which might outweigh the interests of animals. Human interests come first, because, we think, humans are so much *better* than animals. We tend to the Cartesian view, that humans alone have intellectual, moral and spiritual capacities.

Our attitude toward the non-animal world — plants, trees, rocks, rivers, mountains — has also been exploitative. The world has been viewed as an inexhaustible natural resource, to be used up as rapidly as we choose. Conservation policies arise from a belated recognition that resources are limited and that unwise use of resources leads to environmental degradation. In contrast, policies of preservation — the setting aside of areas to be free from adverse human impact — suggest a view of nature as valuable in itself, not merely as a resource to be used.

The speciesist regards the rest of nature as (at best) moral patients, not as agents. There is no moral community with nature. Speciesists may regard nature as a commodity to be consumed, a candy store to be looted, a treasure trove of resources to be conserved or managed on a maximum sustainable yield basis, a garden to be tended, or even a wonderland to be preserved. But, once we have resolved how humans should treat *each other*, the only remaining ethical question is: How should we (humans) treat it (nature)?

the search for an environmental ethic

Even though we in the West are speciesists, our speciesism is certainly more "benign" than that of our ancestors. The industries which use animals try to promote their practices as harmless to animals only because callousness is unacceptable. This suggests that our concept of the moral community is widening. Books such as *Uncle Tom's Cabin* and movies such as *Green Pastures* strike us today as intolerably patronizing. But the concern inherent in these "benignly racist" works in fact marked an important step forward in attitudes held by the majority of whites. Before whites could accept blacks as equals, it was surely necessary to see them as humans with interests, to whom whites had obligations, and not merely as sub-human chattels. Similarly, an "enlightened" benign speciesist ethics is better than an exploitative one, and may mark a stage of development. Indeed it is possible even within a speciesist ethic to develop a concern for the rest of nature. The Amish, for example, believe that

> ... it is sinful for people to misuse or destroy what they did not make. The creation is a unique, irreplaceable gift, and therefore to be used with humility, respect, and skill.

According to Aldo Leopold, the next stage of ethical development is an *environmental ethic*, or as he calls it, the *land ethic*.[6] (See page 43.)

The land ethic simply enlarges the boundaries of the community to include soils, waters, plants and animals, or collectively the land.

To develop an environmental ethic, we will have to change our perceptions of and our behavior towards the rest of nature. Leopold writes,

> ... a land ethic changes the role of *Homo sapiens* from conqueror of the land-community to plain member and citizen of it. It implies respect for his fellow-members, and also respect for the community as such.

Its central value is that

> a thing is right when it tends to preserve the integrity, stability, and beauty of the biotic community. It is wrong when it tends otherwise.

Leopold's concept of nature clearly *includes* humans.

It is a curious irony that at both extremes of the environmental debate we find a common aberration; both wilderness purists[7] and advocates of unrestricted development see nature as separate and different from humans. This concept of nature as essentially *other* is found in numerous writings. For example:

> Nature is coming increasingly under control as a result of restored human confidence and power. We are therefore the first age which can aspire to be free from the tyranny of physical nature that has plagued man since his beginnings.[8]

> The glory of this territory (Yellowstone) is its sublime solitude. Civilization is so universal that man can only see nature in her majesty and primal glory, as it were, in these as yet virgin regions.[9]

This idea that nature does not include humans is supported by the number of commercially available products, from beer to insecticides, that are advertised as "natural." Arsenic is natural, while DDT is not, according to Madison Avenue. Whatever man has concocted is not "natural," and thus one may conclude that man's civilization on the earth is not "natural."[10]

In contrast, we are sympathetic to the concept of nature presented by M. Murie:[11]

> My main concern ... is to suggest a rapprochement with nature in which nature is respected but we people are not required to hate our presence on the planet. I define "natural environment" as any place where organisms exist in mutual relations that are free of direct management or other drastic human intervention; but presence or absence of people is not a criterion.

The first stage in developing an environmental ethic is to liberate ourselves from speciesist beliefs about non-human nature. We might begin by noting that some traditional and contemporary ethical theories could be extended to include animals, at least, and perhaps plants and inanimate natural features too.

A common form of argument in ethical contexts is to try to show that the grounds or reasons alleged to justify a principle would actually justify more than the advocate of that principle had imagined. For example, if someone argues that abortion is always wrong, because it is always wrong to take human life, then he or she must accept that killing in self defense, war, and capital punishment are also wrong, since they also involve the taking of human life. Even a narrower "pro life" principle, such as it is always wrong to take *innocent* human life, would seem to rule out modern warfare, because the killing of non-combatants is inevitable.

The arguments discussed in the next few paragraphs are of this sort. They are designed to show that ethical theories purporting to give an account of our obligations to *humans* also imply that we have obligations towards *non-humans*.

are humans special?

Some philosophers and behavioral scientists have questioned our assumptions that humans, uniquely, possess certain valuable attributes. Some have claimed that dolphins and whales are as intelligent as humans. Well known studies of primate behavior suggest that chimpanzees, at least, are capable of learning and using language, of solving quite complex problems, and of enjoying social relationships.[12] Elephants and dogs sometimes behave in ways which we would describe as altruistic or even heroic if performed by humans,[13] and one writer argues that, behaviorally at least, many animals are as "ethical" as many humans.[14] Another philosopher believes that to be morally significant a being need only have awareness of itself as a separate being, and that many vertebrates (though not fetuses and newborn infants) meet the criterion.[15]

If these animals are relevantly similar to us, then we ought to treat them as we treat humans. If the basis of human rights is that humans are rational or moral, then we ought to ascribe rights to *any* beings which are rational and moral.

The problem with this approach is its rather limited scope. For example, to concentrate on the abilities of a small number of species in language use is to ignore the great mass of animals which do *not* appear to have a language, to reason in sophisticated ways, or to act altruistically. This is a dead end approach, we suggest, and will make little difference in initiating a permanent alteration of environmental attitudes. Perhaps there is a parallel with well-meaning but patronizing claims that artifactual animals such as trick-performing poodles are "almost human," much as non-white tourist visitors to South Africa may be (insultingly) placed in the temporary category of "honorary whites." A useful environmental ethic needs to have a deeper base than merely bringing a few "almost human" animals into the moral community.

utilitarianism and an environmental ethic

The Australian philosopher Peter Singer interprets utilitarianism* as giving moral status to animals.[16] Singer argues that it is the *ability to suffer* which gives a being moral significance — that suffering is bad, no matter what the species of the being. Since all animals above the level of clams are capable of suffering, we should cease to exploit animals, for food, experimentation and product testing. We should also stop destroying and polluting the habitats of wild animals. Singer does not claim that *all* animals are equal in *all* respects, but he does argue that all are equal in one significant respect — the ability to suffer. To deny that they suffer is to refuse to accept overwhelming evidence. To deny that their suffering ought to count is mere prejudice. Singer argues that if we agree that sexism and racism (or slavery) are wrong, we cannot then accept speciesism.

Singer's argument (which is much more detailed than this bare summary suggests) has the merit of appealing to a widely popular and respectable ethical theory, utilitarianism. He does not ask us to make profound revisions to our conceptual framework — merely to acknowledge scientifically based facts, and to extend utilitarianism accordingly. But his theory is vulnerable to the general objections to utilitarianism (see Appendix A.2). Moreover, a utilitarian *environmental ethic* seems impossible, since only sentient beings can be viewed as having moral significance. Non-sentient beings such as plants and inanimate natural objects can have value only as *resources* to sustain sentient beings. To regard only sentient beings as having value is still to draw a line between nature and the human moral community, though at a different place from the speciesist. One author, admittedly with satiric intent, has dubbed the animal liberation movement "fauna chauvinism."

---

*Utilitarianism is one form of ethical thinking, or an ethical theory, which in its simplest form tries to calculate the good that various alternatives achieve, and then choose the alternatives that represent the greatest good. Most engineers, incidentally, are utilitarians. A discussion of utilitarianism and other ethical theories is found in Appendix A.2.

the intrinsic value of nature

We note earlier that some animals do actually meet the criteria, such as rationality, which are supposed to give unique worth to humans. Several philosophers, including Singer and North Carolina State philosopher Tom Regan,[17] have noted that some humans *fail* to meet those criteria, and yet are still regarded as morally significant. Newborn infants, people in irreversible comas, and the severely brain damaged are not rational, autonomous language users, and yet we would not wish *them* to be used for medical experimentation or fattened for the table. Fetuses, which are not even separate beings, are accorded rights by many people. But if humans who *lack* the supposedly special human properties such as reason, concern for others and communicative abilities have moral standing, it is surely inconsistent to refuse to allow that standing to non-humans which *do* possess, in some degree, those properties. Indeed, any animal which comes up to the level of a "human vegetable" ought to be accorded the same consideration. In fact, all the animals we eat, or use in experiments, are more rational than a "human vegetable," and therefore, according to this argument, we ought either to cease our exploitation of animals, or also accept the exploitation of "human vegetables."

This argument, often referred to as the *argument from marginal cases*, has the merit that it can be extended beyond a narrow range of species selected for their similarity to humans. All animals *do* things. Anything with a nervous system can learn about its world, even such limited creatures as flatworms. Regan argues that each animal has its own life to live, and a right to do so. Perhaps this right could extend to all living things. For those who take the trouble to look, there is something of value, something worthy of respect, in the life of every being. In contrast, there might seem little to respect or value in the life of a human with massive brain damage, kept alive only by a respirator. Yet we worry that to turn off the machine would be to "play God," even while we thoughtlessly exploit the lives of other animals and plants as if they had no value at all.

Moreover, each living thing forms part of a biotic community.

The significance of the life of an animal or plant cannot be separated from its ecological context. Living things are prey for each other. Plants and animals, while alive, change the composition of the air, and when dead they enrich the soil. In occupying its ecological niche, a being also contributes to the health and flourishing of the ecosystem of which it is a member. Again, if we are prepared to grant a right to life to a brain-damaged human who contributes nothing, then we ought to recognize the value of other living things, and treat them with respect.

To see value in each living thing, and to recognize that it also has a value because it has an ecological role, is essential to an environmental ethic. It is, perhaps, a short step from there to Leopold's land ethic, seeing value in whole biotic communities or ecosystems, valuing forests, lakes, and mountains not because of their possible utility for humans, but because they are flourishing natural communities, full of creatures living their lives and contributing thereby to the overall health of the community.

universalizability and an environmental ethic

Finally, consider the application of the "universalizability criterion." Singer notes that "Ethics takes a universal point of view."[18] Not only does this require us to go beyond appeals to self interest or sectional interest; it also requires us to put ourselves, in imagination, in the position of others and consider how we would feel if an ethical principle were applied to us.

Speciesists believe that their possession of special qualities entitles them to exploit the rest of nature. The principle appealed to seems to be something like this: A being with greatly superior attributes is entitled to treat beings which lack those attributes as a means to its ends. It would seem to follow that if we humans were in the position of the "inferior," it would be all right for us to be exploited. Suppose, for example, that the earth were invaded by beings from another galaxy whose abilities were enormously greater than ours, in degree and in kind. Perhaps they would have the ability to assimilate, classify and draw upon data in a way superior even to our largest computers. They might have highly developed telepathic or telekinetic powers, and be able to concentrate on an indefinite number of ideas or concepts at the same time. We would be forced to acknowledge that *by our own standards of excellence* these beings were superior to us. Yet we personally would *not* want them to enslave us, to use us for food or experiments to satisfy their scientific curiosity, to display us in zoos, use us for sport, or treat us as pests to be eliminated. Indeed, we would believe that it was their duty *not* to exploit us or to ignore our well-being and interests. Applying the Golden Rule, then, we should not claim the right to exploit and destroy animals or plants needlessly, and abandon speciesism as an ethical principle.

Environmental ethics is a young and highly volatile field, and new ideas and concepts are constantly being advanced. It is difficult to overestimate what the idea of environmental ethics has done to the ethics profession. It has, in the space of a few years, broadened the spectrum of what one should consider in an ethical context. This widening of our moral community is a tremendous challenge to the profession of ethics, and accordingly, the field of environmental ethics is in a rapid state of development.

In the previous chapter we discuss the difficulties in defining professional ethics and what is meant by ethical professional conduct. This field, just like environmental ethics, is undergoing rapid development.

It is therefore with some trepidation that in the next chapter we turn to the problem of how professional engineering ethics and environmental ethics blend together for engineers engaged in work relating to nature, natural resources, and the environment. In engineering parlance, there are an awful lot of variables loose. Nevertheless, the infusion of environmental ethics into professional conduct is a major new force in engineering.

references

1. Stevenson, C.C. *Ethics and Language*, Yale University Press, New Haven, CT, 1944.
2. Katzenstein, A.W. *An Updated Perspective on Acid Rain*, Edi-

son Electric Institute, New York, 1981.

3. White, L. Jr. "The Historical Roots of Our Ecologic Crisis," *Science* 155, 1203-1207, 10 March 1967.

4. Montefiore, H. *Can Man Survive?* Fontana Books, London, 1970, as quoted in Passmore, J., *Man's Responsibility for Nature*, Charles Scribner's Sons, New York, 1974.

5. Berry, W. *The Unsettling of America*, Avon Books, New York, 1977.

6. Leopold, A. "The Land Ethic," in *A Sand County Almanac with Essays on Conservation from Round River*, Oxford University Press, New York, 1966.

7. Graber, L.H. *Wilderness as Sacred Space*, American Society of Geographers, Washington DC, 1976.

8. Methesne, E.G. "Technology and Religion," *Theology Today* 23, 1965.

9. Nash,R. *Wilderness and the American Mind*, Yale University Press, New Haven, CT, 1973.

10. Vesilind, P.A. discussion of "Can and Ought We to Follow Nature?" *Environmental Ethics* 1,4, 1979.

11. Murie, M. "Evaluation of Natural Environments" in *Indicators of Environmental Quality*, W.A. Thomas (Ed.), Plenum Press, New York, 1972.

12. Linden, E. *Apes,Men and Language*, Saturday Review Press/ E.P.Dutton, New York, 1975.

13. Midgley, M. *Beast and Man*, The Harvester Press, Brighton, U.K., 1979.

14. MacIver, A.M. "Ethics and the Beetle," *Analysis*, 8, 65-70, 1948.

15. Tooley, M. "Abortion and Infanticide" *Philosophy and Public Affairs*, 2, 37-65, 1972.

16. Singer, P. *Animal Liberation*, New York Review, New York, 1975.

17. Regan, T. *The Case for Animal Rights*, University of California Press, Berkely CA, 1983.

18. Singer, P. *Practical Ethics*, Cambridge University Press, New York, 1979.

1.3 Environmental Ethics and Professional Engineering

IN THE FIRST section of this book the idea of ethics is introduced, and it is argued that ethics has for thousands of years been thought of as a framework for organized thinking of how one relates to other people. Where decisions have to be made and values conflict, ethics is a systematic approach to providing guidance for the resolution of such conflicts. In engineering, as in other professions, value-conflicts can occur on the job, and ethical decisions must be made by the engineer. While the "Code of Ethics" provides some guidance, and while laws further define the limits of conduct, there remain many situations where personal values must be put into play in making engineering decisions.

In the second section of this book, the idea of environmental ethics is introduced, and it is argued that the moral community should encompass not only other people, but the world generally. The idea of animals and plants, and even inanimate objects having a right to exist, or a "standing," is put forth. This is a very special form of ethics, broadly labeled environmental ethics.

What remains is to combine the concept of environmental ethics with professional engineering ethics, and to show how the recognition of an expanded moral community can affect engineering decisions.

engineers and environmental concerns

Scattered throughout the text above are examples of where engineering decisions affect not only people but other parts of nature as well. The importance of this concern in the practice of engineering is a very new idea, since engineering has often been advertised as a "people serving profession." Nowhere is engineering claimed to be "nature serving." Engineers work *for* other people, and have special responsibility toward their employers and clients. And their employers and clients are invariably people. But who, as in Dr. Seuss' *Lorax*, speaks for the trees? Who is the Mr. Onceler who tries to warn us of stupid and unthinking acts which are devastating to our environment?

One approach is to allow the "environmentalists" to do their thing and keep an eye on the world. Most industries are scared to death of the much-storied "little old lady in tennis shoes." And engineers might be tempted to abrogate their responsibility for environmental decisions by suggesting that unless someone is vehemently opposed to their actions, then the actions must be environmentally sound. This of course is a classical cop-out.

Some engineers use the related technique of "public hearing" to place a stamp of approval on their plans. This method has been perfected by the Corps of Engineers. The technique works as follows: First, an engineering decision is made to construct some large facility, then the people who would benefit from this facility such as real estate developers and local merchants are asked to comment. Public hearings are often stacked (not by any malicious purpose) in favor of those who would attain some immediate gain from the expenditure of public money. Transcripts of such hearings reveal how, once a project is initiated, it can gain momentum and attain a life of its own, quite apart from any rational reasoning for its continuation.

One example of many such engineering projects is the "strange ditch in the Everglades."[1] In concert with real estate developers and farmers, the Corps constructed a large ditch to drain a freshwater marshland near Lake Okeechobee in the Florida Everglades. In none of the planning documents was it noted that this drainage would allow severe saltwater intrusion and would wipe out a large fraction of many rare and endangered birds such as the wood stork, the great white heron, and the bald eagle.[2] All available disinterested engineering and hydrologic skill cautioned against the drainage ditch, but the Corps proceeded anyway, inflating the benefits by increasing the land values as needed to cover the escalating costs of construction. The project had no apparent worth, and caused considerable harm, but had to be completed because of the tremendous momentum it had gained. Where were the engineers with courage to ask the right questions?

the engineers and environmental impact statements

Due in part to the inability of engineers to assimilate the effect of their projects on other species and nature generally, the Congress passed and President Nixon signed into law on 1 January 1970 the National Environmental Policy Act (NEPA). This legislation, in tandem with the Endangered Species Act passed four year earlier, provided the ammunition for environmental groups to battle the designs of government projects. The chief weapon was a little-discussed section of NEPA, Section 102, which mandated that any government project which significantly affects the environment, must have an *Environmental Impact Statement*, or an EIS. The purpose of this statement was to *force* the engineers, planners and managers to at least *consider* the effect their plans would have on other species, on the visual environment, and on long term conservation and use of resources.

At first engineers pooh-poohed the concept. They figured they would write some generic disclaimer and attach it to every report, just like a job discrimination notice is presently handled. But they were foiled by the courts. For example, the New Hope Dam Project (later the B. Everett Jordan Dam), which is now costing well over $100 million, was first graced with a mere 17 page EIS. It did not take the court long to determine that this was totally inadequate. The Corps was forced to reconsider the question and to develop a comprehensive statement outlining the impact of this project on the environment. (The resolution of this conflict is described in an article starting on page 101.)

Over the years, the form of the EIS has changed, and the procedures have been streamlined, but the effect remains. It is no longer possible, as was done in the Everglades project, to simply ignore the bald eagles, for example. Their survival was then not considered an engineering problem. Now, with the EIS, the welfare of other species becomes a part of the planning process.

We do not mean to imply that the EIS has infused environmental ethics into the engineering profession, and now the problem is solved. While the EIS helps a great deal in forcing engineers to expand their thinking on environmental problems, it is not a total solution. There are several reasons for this:

1. The EIS is applicable to only *some* federal projects. Most large projects by the Department of Defense, like weapons testing for example, do not require EISs.

2. The EIS is written only for projects that *significantly* affect the environment. The determination as to what is and is not significant is left to the agency to decide.

3. While most states have some form of EIS, most of these are poorly enforced. Thus few engineering projects funded by the states have this control.

4. There is no EIS or equivalent for all projects by private developers. Thus if a large agricultural firm wants to strip the surface of a huge peat deposit in order to carry on high-intensity farming (with almost certain contamination of the groundwater), there is no mechanism (other than some permits) which will force the engineers and entrepreneurs to consider the effects on the environment.

5. There is no mechanism presently available for making the value judgments necessary in the conclusion of environmental impact statements.

Consider this last point for a moment. Suppose the Corps finds that flooding a marshland would most probably cause the destruction of habitat for some bird, say the bald eagle. In the EIS, the pros and cons of the project are listed, including alternatives to the project, and at the end, a decision is to be made. "Yes" the project is to go forward, or "no," it is to be stopped. Who decides that, and how?

It is easy to calculate the *positive* effect of a project by just estimating the increase in property value or some other benefit. But what about cost? What does a bald eagle cost? How does one place a value on such a bird? And yet, the engineers want the EIS to be quantitative so that a clear decision can be made.

What happens is that the engineers (mostly) arrange their calculations of environmental effect in such a way that the outcome reflects *exactly what they wanted to do in the first place*. Since there are no standard methods for writing the statements, there are many ways to manipulate the judgmental data in order to obtain the desired result.

The most important criticism of the EIS, however is this:

6. Even though the EIS might show severe damage to the environment, and even though everyone (including the engineers) agrees that this is the case, there is no guarantee that the project will be scrapped. In many cases, that decision is made in the White House, where political pressures can play an overriding role. This point is important, so it deserves repeating: The EIS, even though it might be highly negative, can not automatically stop or even change a project. The decision to do so is reserved for the political process.

If this is true, why then worry the engineers with concerns of values and ethics? Are not engineers then simply hired hands, doing the bidding of the society which employs them?

The response to that question is that engineers *do* decide policy in many ways. On a local and mundane level, engineers are making seemingly minor decisions daily which affect all of us. For example, the routing of a local highway across a stream which may be a rare habitat for the endangered venus fly trap is an engineering decision that will get no press whatever, much less receive White House consideration. And on a national level, the mere initiation of a major project will result in a momentum which will be difficult to reverse. Thus it is not enough simply to defer engineering decisions to public comment or environmental oversight. The *engineers* need to introduce environmental concerns into the planning of projects before the plans become public documents. The decisions of the engineers *do* make a difference.

making the difference

Clearly, engineers are in a unique position to make a difference. Private and governmental clients, if they are to get things done, *must* use engineering knowhow. They do not necessarily have to use other experts such as sociologists, philosophers, epidemiologists, or even planners, but they *must* hire engineers. Thus the engineer has the responsibility of viewing his/her role in the broadest sense — to introduce alternatives, concepts and values which the client may otherwise never consider. For example, if a city hires an engineering firm to design a highway, it is possible for the engineer to ignore all the aspects of the job except the determination of the best (safest and least expensive) highway alignment. What we are suggesting is that the engineer, recognizing that there are many conflicting values involved in such a project, and that it is unlikely that the municipality will seek expert assistance in the resolution of such conflicts, has a responsibility to introduce the value (ethical) questions in the report to the client.

For example, a New Zealand engineer designing a gas pipeline which would go through a burial ground sacred to a Maori tribe should introduce the conflict associated with the use of such land for a pipeline into the design process. Just because the alignment of the pipeline would be least expensive if placed across the sacred (tapu) mountain, such a decision may not properly serve society. The engineer should recognize these conflicts in the early design and introduce solutions which minimize the unwanted side-effects of engineering projects.

But it is not possible to sensitize engineers to broader social values without an education which introduces them to concepts of religion, sociology, political science, anthropology, history, and perhaps most importantly, ethics. It is imperative that engineers ask the proper questions about how their work will affect others, and the formulation of these questions is only possible through a broad engineering education.

Engineering education should address not only technical questions, but introduce the concept of values, and the methodologies for decision-making when values conflict. Ethical issues always arise on any engineering job, and the engineer must consider alternatives, recognize social costs and benefits, be aware of tradeoff possibilities, and understand the limits of technical solutions and economic analyses.

A truly professional engineer will infuse ethics into his/her decision-making, and with the increasing pressure on the natural environment, a growing population, and accelerated technological development, environmental ethics will play an ever greater role in the engineer's role in society.

references

1. Laycock, G. *The Diligent Destroyers*, Ballantine Books, New York, 1970.
2. Douglas, W.O. "The Corps of Engineers: The Public Be Damned," *Playboy* July, 1969.

Part II
Readings and Case Studies

2.1 The Land Ethic
Aldo Leopold

Aldo Leopold (1887-1948), a naturalist and writer, produced perhaps the first formal statement on an environmental ethic. His A Sand County Almanac *is full of touching and vivid observations of the natural environment. He was a realist, and recognized that the preservationists' ideas could never be implemented in their pure form, and yet he also rejected the ideas of the early conservationists, who valued nature only for what it could provide in the form of material wealth. He insisted that nature has an intrinsic value beyond that of dollars and cents, and that this value could not necessarily be expressed in terms of a benefit-cost ratio.*

In this short excerpt from his book, the idea of the land ethic is first introduced. Bear in mind in reading this article that it was written in the 1930s, and that the idea of ethics applied to land, thus widening the moral community, was a revolutionary concept.

WHEN GOD-LIKE Odysseus returned from the wars in Troy, he hanged all on one rope a dozen slave girls of his household who he suspected of misbehavior during his absence.

This hanging involved no question of propriety. The girls were property. The disposal of property was then, as now, a matter of expediency, not of right and wrong.

Concepts of right and wrong were not lacking from Odysseus' Greece: witness the fidelity of his wife through the long years

Excerpted from Leopold, A. *A Sand County Almanac*, Oxford University Press, New York, 1966. Used with permission.

before at last his black prowed galleys clove the wine-dark seas for home. The ethical structure of that day covered wives, but had not yet been extended to human chattels. During the three thousand years which have since elapsed, ethical criteria have been extended to many fields of conduct, with corresponding shrinkages in those judged by expediency only.

This extension of ethics, so far studied only by philosophers, is actually a process in ecological evolution. Its sequences may be described in ecological as well as in philosophical terms. An ethic, ecologically, is a limitation on freedom of action in the struggle for existence. An ethic, philosophically, is a differentiation of social from anti-social conduct. These are two definitions of one thing. The thing has its origin in the tendency of interdependent individuals or groups to evolve modes of co-operation. The ecologist calls these symbioses. Politics and economics are advanced symbioses in which the original free-for-all competition has been replaced, in part, by co-operative mechanisms with an ethical content.

The complexity of co-operative mechanisms has increased with population density, and with the efficiency of tools. It was simpler, for example, to define the anti-social uses of sticks and stones in the days of the mastodons than of bullets and billboards in the age of motors.

The first ethics dealt with the relation between individuals; the Mosaic Decalogue is an example. Later accretions dealt with the relation between the individual and society. The Golden Rule tries to integrate the individual to society; democracy to integrate social organization to the individual.

There is as yet no ethic dealing with man's relation to land and to the animals and plants which grow upon it. Land, like Odysseus' slave girls, is still property. The land-relation is still strictly economic, entailing privileges but not obligations.

The extension of ethics to this third element in human environment is, if I read the evidence correctly, an evolutionary possibility and an ecological necessity. It is the third step in a sequence. The first two have already been taken. Individual thinkers since the days of Ezekiel and Isaiah have asserted that the despoilation of land is not only inexpedient but wrong. Society, however, has not yet affirmed their belief. I regard the present conservation movement as the embryo of such an affirmation.

An ethic may be regarded as a mode of guidance for meeting ecological situations so new or intricate or involving such deferred reactions, that the path of social expediency is not discernible to the average individual. Animal instincts are modes of guidance for the individual in meeting such situations. Ethics are possibly a kind of community instinct in-the-making.

the community concept

All ethics so far evolved rest upon a single premise: that the individual is a member of a community of interdependent parts. His instincts prompt him to compete for his place in the community, but his ethics prompt him also to co-operate (perhaps in order that there may be a place to compete for).

The land ethic simply enlarges the boundaries of the community to include soils, waters, plants, and animals, or collectively: the land.

This sounds simple: do we not already sing our love for and obligation to the land of the free and the home of the brave? Yes, but just what and whom do we love? Certainly not the soil, which we are sending helter-skelter downriver. Certainly not the waters, which we assume have no function except to turn turbines, float barges, and carry off sewage. Certainly not the plants, of which we exterminate whole communities without batting an eye. Certainly not the animals, of which we have already extirpated many of the largest and most beautiful species. A land ethic of course cannot prevent the alteration, management, and use of these 'resources', but it does affirm their right to continued existence, and at least in spots, their continued existence in a natural state.

In short, a land ethic changes the role of *Homo sapiens* from conqueror of the land-community to plain member and citizen of it. It implies respect for his fellow-members, and also respect for the community as such.

In human history, we have learned (I hope) that the conqueror role is eventually self-defeating. Why? Because it is implicit in such a role that the conqueror knows, *ex cathedra*, just what makes the community clock tick, and just what and who is valuable, and what and who is worthless, in community life. It always turns out that he knows neither, and this is why his conquests eventually defeat themselves.

In the biotic community, a parallel situation exists. Abraham knew exactly what the land was for — it was to drip milk and honey into Abraham's mouth. At the present moment, the assurance with which we regard this assumption is inverse to the degree of our education.

The ordinary citizen today assumes that science knows what makes the community clock tick; the scientist is equally sure that he does not. He knows that the biotic mechanism is so complex that its workings may never be fully understood.

That man is, in fact, only a member of a biotic team is shown by an ecological interpretation of history. Many historical events, hitherto explained solely in terms of human enterprise, were actually biotic interactions between people and land. The characteristics of the land determined the facts quite as potently as the characteristics of the men who lived on it.

Consider, for example, the settlement of the Mississippi valley. In the years following the Revolution, three groups were contending for its control — the native Indian, the French and English traders, and the American settlers. Historians wonder what would have happened if the English at Detroit had thrown a little more weight into the Indian side of those tipsy scales which decided the outcome of the colonial migration into the cane-lands of Kentucky. It is time now to ponder the fact that the cane-lands, when subjected to the particular mixture of forces represented by the cow, plow, fire, and axe of the pioneer, became bluegrass. What if the plant succession inherent in this dark and bloody ground had, under the impact of these forces, given us some worthless sedge, shrub, or weed? Would Boone and Kenton have held out? Would there have been any overflow into Ohio, Indiana, Illinois, and Missouri? Any Louisiana Purchase? Any transactional union of new states? Any Civil War?

Kentucky was one sentence in the drama of history. We are

commonly told what the human actors in this drama tried to do, but we are seldom told that their success, or the lack of it, hung in large degree on the reaction of particular soils to the impact of the particular forces exerted by their occupancy. In the case of Kentucky, we do not even know where the bluegrass came from — whether it is a native species, or a stowaway from Europe.

Contrast the cane-lands with what hindsight tells us about the Southwest, where pioneers were equally brave, resourceful, and persevering. The impact of occupancy here brought no bluegrass, or other plant fitted to withstand the bumps and buffetings of hard use. This region, when grazed by livestock, reverted through a series of more and more worthless grasses, shrubs, and weeds to a condition of unstable equilibrium. Each recession of plant types bred erosion; each increment to erosion bred a further recession of plants. The result today is a progressive and mutual deterioration, not only of plants and soils, but of the animal community subsisting thereon. The early settlers did not expect this: on the cienegas of New Mexico some even cut ditches to hasten it. So subtle has been its progress that few residents of the region are aware of it. It is quite invisible to the tourist who finds this wrecked landscape colorful and charming (as indeed it is, but it bears scant resemblance to what it was in 1848).

This same landscape was "developed" once before, but with quite different results. The Pueblo Indians settled the Southwest in pre-Columbian times, but they happened *not* to be equipped with range livestock. Their civilization expired, but not because their land expired.

In India, regions devoid of any sod-farming grass have been settled, apparently without wrecking the land, by the simple expedient of carrying the grass to the cow, rather than vice versa. (Was this the result of some deep wisdom, or was it just good luck? I do not know.)

In short, the plant succession steered the course of history; the pioneer simply demonstrated, for good or ill, what successions inhered in the land. Is history taught in this spirit? It will be, once the concept of land as a community really penetrates our intellectual life.

the ecological conscience

Conservation is a state of harmony between men and land. Despite nearly a century of propaganda, conservation still proceeds at a snail's pace; progress still consists largely of letterhead pieties and convention oratory. On the back forty we still slip two steps backward for each forward stride.

The usual answer to this dilemma is 'more conservation education.' No one will debate this, but is it certain that only the *volume* of education needs stepping up? Is something lacking in the *content* as well?

It is difficult to give a fair summary of its content in brief form, but, as I understand it, the content is substantially this: obey the law, vote right, join some organizations, and practice what conservation is profitable on your own land; the government will do the rest.

Is not this formula too easy to accomplish anything worthwhile? It defines no right or wrong, assigns no obligation, calls for no sacrifice, implies no change in the current philosophy of values. In respect of land-use, it urges only enlightened self-interest. Just how far will such education take us? An example will perhaps yield a partial answer.

By 1930 it had become clear to all except the ecologically blind that southwestern Wisconsin's topsoil was slipping seaward. In 1933 the farmers were told that if they would adopt certain remedial practices for five years, the public would donate CCC labor to install them, plus necessary machinery and materials. The offer was widely accepted, but the practices were widely forgotten when the five-year contract period was up. The farmers continued only those practices that yielded an immediate and visible economic gain for themselves.

This led to the idea that maybe farmers would learn more quickly if they themselves wrote the rules. Accordingly the Wisconsin Legislature in 1937 passed the Soil Conservation District Law. This said to farmers, in effect: *We, the public, will furnish you free technical service and loan you specialized machinery, if you will write your own rules for land-use. Each county may write its own rules, and these will have the force of law.* Nearly all the counties promptly organized to accept the proffered help, but after a decade of operation, *no county has yet written a single rule.* There has been visible progress in such practices as strip-cropping, pasture renovation, and soil liming, but none in fencing woodlots against grazing, and none in excluding plow and cow from steep slopes. The farmers, in short, have selected those remedial practices which were profitable anyhow, and ignored those which were profitable to the community, but not clearly profitable to themselves.

When one asks why no rules have been written one is told that the community is not yet ready to support them; education must precede rules. But the education actually in progress makes no mention of obligations to land over and above those dictated by self-interest. The net result is that we have more education but less soil, fewer healthy woods, and as many floods as in 1937.

The puzzling aspect of such situations is that the existence of obligations over and above self-interest is taken for granted in such rural community enterprises as the betterment of roads, schools, churches, and baseball teams. Their existence is not taken for granted, nor as yet seriously discussed, in bettering the behavior of the water that falls on the land, or in the preserving of the beauty or diversity of the farm landscape. Land-use ethics are still governed wholly by economic self-interest, just as social ethics were a century ago.

To sum up: we asked the farmer to do what he conveniently could to save his soil, and he has done just that, and only that. The farmer who clears the woods off a 75 per cent slope, turns his cows into the clearing, and dumps its rainfall, rocks, and soil into the community creek, is still (if otherwise decent) a respected member of society. If he puts lime on his fields and plants his crops on contour, he is still entitled to all the privileges and emoluments of his Soil Conservation District. The District is a beautiful piece of social machinery, but it is coughing along on two cylinders because we have been too timid, and too anxious for quick success, to tell the farmer the true magnitude of his obligations. Obligations have no meaning without conscience, and the problem we face is the extension of the social conscience from people to land.

No important change in ethics was ever accomplished without an internal change in our intellectual emphasis, loyalties, affections, and convictions. The proof that conservation has not yet touched these foundations of conduct lies in the fact that philosophy and religion have not yet heard of it. In our attempt to make conservation easy, we have made it trivial.

substitutes for a land ethic

When the logic of history hungers for bread and we hand out a stone, we are at pains to explain how much the stone resembles bread. I now describe some of the stones which serve in lieu of a land ethic.

One basic weakness is a conservation system based wholly on economic value. Wildflowers and songbirds are examples. Of the 22,000 higher plants and animals native to Wisconsin, it is doubtful whether more than 5 per cent can be sold, fed, eaten, or otherwise put to economic use. Yet these creatures are members of the biotic community, and if (as I believe) its stability depends on its integrity, they are entitled to continuance.

When one of these non-economic categories is threatened, and if we happen to love it, we invent subterfuges to give it economic importance. At the beginning of the century songbirds were supposed to be disappearing. Ornithologists jumped to the rescue with some distinctly shaky evidence to the effect that insects would eat us up if birds failed to control them. The evidence had to be economic in order to be valid.

It is painful to read these circumlocutions today. We have no land ethic yet, but we have at least drawn nearer the point of admitting that birds should continue as a matter of biotic right, regardless of the presence or absence of economic advantage to us.

A parallel situation exists in respect of predatory mammals, raptorial birds, and fish-eating birds. Time was when biologists somewhat overworked the evidence that these creatures preserve the health of game by killing weaklings, or that they control rodents for the farmer, or that they prey only on "worthless" species. Here again, the evidence had to be economic in order to be valid. It is only in recent years that we hear the more honest argument that predators are members of the community, and that no special interest has the right to exterminate them for the sake of a benefit, real or fancied, to itself. Unfortunately this enlightened view is still in the talk stage. In the field the extermination of predators goes merrily on: witness the impending erasure of the timber wolf by fiat of Congress, the Conservation Bureaus, and many state legislatures.

Some species of trees have been "read out of the party" by economics-minded foresters because they grow too slowly, or have too low a sale value to pay as timber crops: white cedar, tamarack, cypress, beech, and hemlock are examples. In Europe, where forestry is ecologically more advanced, the non-commercial tree species are recognized as members of the native forest community, to be preserved as such, within reason. Moreover some (like beech) have been found to have a valuable function in building up soil fertility. The interdependence of the forest and its constituent tree species, ground flora, and fauna is taken for granted.

Lack of economic value is sometimes a character not only of species or groups, but of entire biotic communities: marshes, bogs, dunes and "deserts" are examples. Our formula in such cases is to relegate their conservation to government as refuges, monuments, or parks. The difficulty is that these communities are usually interspersed with more valuable private lands; the government cannot possibly own or control such scattered parcels. The net effect is that we have relegated some of them to ultimate extinction over large areas. If the private owner were ecologically minded, he would be proud to be the custodian of a reasonable proportion of such areas, which add diversity and beauty to his farm and to his community.

In some instances, the assumed lack of profit in these "waste" areas has proved to be wrong, but only after most of them had been done away with. The present scramble to reflood muskrat marshes is a case in point.

There is a clear tendency in American conservation to relegate to government all necessary jobs that private landowners fail to perform. Government ownership, operation, subsidy, or regulation is now widely prevalent in forestry, range management, soil and watershed management, park and wilderness conservation, fisheries management, and migratory bird management, with more to come. Most of this growth in governmental conservation is proper and logical, some of it is inevitable. That I imply no disapproval of it is implicit in the fact that I have spent most of my life working for it. Nevertheless the question arises: What is the ultimate magnitude of the enterprise? Will the tax base carry its eventual ramifications? At what point will governmental conservation, like the mastodon, become handicapped by its own dimensions? The answer, if there is any, seems to be in a land ethic, or some other force which assigns more obligation to the private landowner.

Industrial landowners and users, especially lumbermen and stockmen, are inclined to wail long and loudly about the extension of government ownership and regulation to land, but (with notable exceptions) they show little disposition to develop the only visible alternative: the voluntary practice of conservation on their own lands.

When the private landowner is asked to perform some unprofitable act for the good of the community, he today assents only with outstretched palm. If the act costs him cash this is fair and proper, but when it costs only fore-thought, open-mindedness, or time, the issue is at least debatable. The overwhelming growth of land-use subsidies in recent years must be ascribed, in large part, to the government's own agencies for conservation education: the land bureaus, the agricultural colleges, and the extension services. As far as I can detect, no ethical obligation toward land is taught in these institutions.

To sum up: a system of conservation based solely on economic self-interest is hopelessly lopsided. It tends to ignore, and thus eventually to eliminate, many elements in the land community that lack commercial value, but that are (as far as we know) essential to its healthy functioning. It assumes, falsely, I think, that the economic parts of the biotic clock will function without the uneconomic parts. It tends to relegate to government many functions eventually too large, too complex, or too widely dispersed to be performed by government.

An ethical obligation on the part of the private owner is the only visible remedy for these situations.

2.2 The Tragedy of the Commons

Garrett Hardin

Garrett Hardin (1915-) has been called "one of the intellectual leaders of our time" by Science *magazine. Yet he probably would not think of himself as that, because he has been, since the tumultuous 1960s, preaching what he considers simple good sense. The tenor of all of his writings is: "OK, now what is the bottom line? What is the unemotional, unvarnished truth in this matter? I don't care who it offends as long as it is the truth." Understandably, he has been criticized and even vilified by numerous writers who decry his inhumanity and insensitivity. But the opposite is true. Hardin writes **about** humanity and all of our problems and foibles. While he may at times appear crass, he is always thought-provoking. He shakes the trees to see what might drop, and on whose head.*

In this piece he starts by discussing the nature of what he calls "the tragedy of the commons," and in the remainder of the article (not reproduced here) he applies this concept to the problem of over-population and birth control. The "tragedy" applies equally well to many other environmental concerns and seems to be at the very root of our problems in achieving a stable, livable environment.

IN ECONOMIC affairs, *The Wealth of Nations* (1776) popularized the "invisible hand," the idea that an individual who "in-

Excerpted from Hardin, G. "Tragedy of the Commons," *Science*, v. 162, 13 December 1968. Used with permission.

tends only his own gain," is, as it were, "led by an invisible hand to promote ... the public interest." Adam Smith did not assert that this was invariably true, and perhaps neither did any of his followers. But he contributed to a dominant tendency of thought that has ever since interfered with positive action based on rational analysis, namely, the tendency to assume that decisions reached individually will, in fact, be the best decisions for an entire society. If this assumption is correct it justifies the continuance of our present policy of laissez-faire in reproduction. If it is correct we can assume that men will control their individual fecundity so as to produce the optimum population. If the assumption is not correct, we need to reexamine our individual freedoms to see which ones are defensible.

tragedy of freedom in a commons

The rebuttal to the invisible hand in population control is to be found in a scenario first sketched in a little-known pamphlet in 1833 by a mathematical amateur named William Forster Lloyd (1794-1852). We may well call it "the tragedy of the commons," using the word "tragedy" as the philosopher Whitehead used it: "The essence of dramatic tragedy is not unhappiness. It resides in the solemnity of the remorseless working of things." He then goes on to say, "This inevitableness of destiny can only be illustrated in terms of human life by incidents which in fact involve unhappiness. For it is only by them that the futility of escape can be made evident in the drama."

The tragedy of the commons develops in this way. Picture a pasture open to all. It is to be expected that each herdsman will try to keep as many cattle as possible on the commons. Such an arrangement may work reasonably satisfactorily for centuries because tribal wars, poaching, and disease keep the numbers of both man and beast well below the carrying capacity of the land. Finally, however, comes the day of reckoning, that is, the day when the long-desired goal of social stability becomes a reality. At this point, the inherent logic of the commons remorselessly generates tragedy.

As a rational being, each herdsman seeks to maximize his gain. Explicitly or implicitly, more or less consciously, he asks, "What is the utility *to me* of adding one more animal to my herd?" This utility has one negative and one positive component.

1) The positive component is a function of the increment of one animal. Since the herdsman receives all the proceeds from the sale of the additional animal, the positive utility is nearly $+1$.

2) The negative component is a function of the additional overgrazing created by one more animal. Since, however, the effects of overgrazing are shared by all the herdsmen, the negative utility for any particular decision-making herdsman is only a fraction of -1.

Adding together the component partial utilities, the rational herdsman concludes that the only sensible course for him to pursue is to add another animal to his herd. And another; and another.... But this is the conclusion reached by each and every rational herdsman sharing a commons. Therein is the tragedy. Each man is locked into a system that compels him to increase his herd without limit — in a world that is limited. Ruin is the destination toward which all men rush, each pursuing his own best interest in a society that believes in the freedom of the commons. Freedom in a commons brings ruin to all.

Some would say that this is a platitude. Would that it were! In a sense, it was learned thousands of years ago, but natural selection favors the forces of psychological denial. The individual benefits as an individual from his ability to deny the truth even though society as a whole, of which he is a part, suffers. Education can counteract the natural tendency to do the wrong thing, but the inexorable succession of generations requires that the basis for this knowledge be constantly refreshed.

A simple incident that occurred a few years ago in Leominster, Massachusetts, shows how perishable the knowledge is. During Christmas shopping season the parking meters downtown were covered with plastic bags that bore tags reading: "Do not open until after Christmas. Free parking courtesy of the mayor and city council.'' In other words, facing the prospect of an increased demand for already scarce space, the city fathers reinstituted the system of the commons. (Cynically, we suggest that they gained more votes than they lost by this retrogressive act.)

In an approximate way, the logic of the commons has been understood for a long time, perhaps since the discovery of agriculture or the invention of private property in real estate. But it is understood mostly in special cases which are not sufficiently generalized. Even at this late date, cattlemen leasing national land on the western ranges demonstrate no more than an ambivalent understanding, in constantly pressuring federal authorities to increase the head count to the point where overgrazing produces erosion and weed-dominance. Likewise, the oceans of the world continue to suffer from the survival of the philosophy of the commons. Maritime nations still respond automatically to the shibboleth of the ''freedom of the seas.'' Professing to believe in the ''inexhaustible resources of the oceans,'' they bring species after species of fish and whales closer to extinction.

The National Parks present an instance of the working out of the tragedy of the commons. At present, they are open to all, without limit. The parks themselves are limited in extent — there is only one Yosemite Valley — whereas population seems to grow without limit. The values that visitors seek in the parks are steadily eroded. Plainly, we must soon cease to treat the parks as commons or they will be of no value to anyone.

What shall we do? We have several options. We might sell them off as private property. We might keep them as public property, but allocate the right to enter them. The allocation might be on the basis of wealth, by use of an auction system. It might be on the basis of merit, as defined by some agreed-upon standards. It might be by lottery. Or it might be on a first-come, first-served basis, administered to long queues. These, I think are all the reasonable possibilities. They are all objectionable. But we must choose — or acquiesce in the destruction of the commons that we call our National Parks.

pollution

In a reverse way, the tragedy of the commons reappears in problems of pollution. Here it is not a question of taking something out of the commons, but of putting something in — sewage, or chemical, radioactive, and heat wastes into water; noxious and dangerous fumes into the air; and distracting and unpleasant advertising signs into the line of sight. The calculations of utility are much the same as before. The rational man finds that his share of the cost of the wastes he discharges into the commons is less than the cost of purifying his wastes before releasing them. Since this is true for everyone, we are locked into a system of "fouling our own nest," so long as we behave only as independent, rational, free-enterprisers.

The tragedy of the commons as a food basket is averted by privated property, or something formally like it. But the air and waters surrounding us cannot readily be fenced, and so the tragedy of the commons as a cesspool must be prevented by different means, like coercive laws or taxing devices that make it cheaper for the polluter to treat his pollutants than to discharge them untreated. We have not progressed as far with the solution of this problem as we have with the first. Indeed, our particular concept of private property, which deters us from exhausting the positive resources of the earth, favors pollution. The owner of a factory on the bank of a stream — whose property extends to the middle of the stream — often has difficulty seeing why it is not his natural right to muddy the waters flowing past his door. The law, always behind the times, requires elaborate stitching and fitting to adapt it to this newly perceived aspect of the commons.

The pollution problem is a consequence of population. It did not much matter how a lonely American frontiersman disposed of his waste. "Flowing water purifies itself every 10 miles," my grandfather used to say, and the myth was near enough to the truth when he was a boy, for there were not too many people. But as population became denser, the natural chemical and biological processes became overloaded, calling for a redefinition of property rights.

how to legislate temperance?

Analysis of the pollution problem as a function of population density uncovers a not generally recognized principle of morality, namely: *the morality of an act as a function of the state of the system at the time it is performed.* Using the commons as a cesspool does not harm the general public under frontier conditions, because there is no public; the same behavior in a metropolis is unbearable. A hundred and fifty years ago a plainsman could kill an American bison, cut out only the tongue for his dinner, and discard the rest of the animal. He was not in an important sense being wasteful. Today, with only a few thousand bison left, we would be appalled at such behavior.

In passing, it is worth noting that the morality of an act cannot be determined from a photograph. One does not know whether a man killing an elephant or setting a fire to the grassland is harming others until one knows the total system in which his act appears. "One picture is worth a thousand words," said an ancient Chinese; but it may take 10,000 words to validate it. It is as tempting to ecologists as it is to reformers in general to try to persuade others by way of the photographic shortcut. But the essence of an argument cannot be photographed; it must be presented rationally — in words.

That morality is system-sensitive escaped the attention of most codifiers of ethics in the past. "Thou shalt not ..." is the form of traditional ethical directives which make no allowance for particular circumstances. The laws of our society follow the pattern of ancient ethics, and therefore are poorly suited to governing a complex, crowded, changeable world. Our epicyclic solution is to augment statutory law with administrative law. Since it is practically impossible to spell out all the conditions under which it is safe to burn trash in the back yard or to run an automobile without smog-control, by law we delegate the details to bureaus. The result is administrative law, which is rightly feared for an ancient reason — *Quis cultodiet ipsos custodes?* — "Who shall watch the watchers themselves?" John Adams said that we must have "a government of laws and not men." Bureau administrators, trying to evaluate the morality of acts in the total system, are singularly liable to corruption, producing a government by men, not laws.

Prohibition is easy to legislate (though not necessarily to enforce); but how do we legislate temperance? Experience indicates that it can be accomplished best through the mediation of administrative law. We limit possibilities unnecessarily if we suppose that the sentiment of *Quis custodiet* denies us the use of administrative law. We should rather retain the phrase as a perpetual reminder of fearful dangers we cannot avoid. The great challenge facing us now is to invent the corrective feedbacks that are needed to keep custodians honest. We must find ways to legitimate the needed authority of both the custodians and the corrective feedbacks.

2.3 The Kepone Tragedy
W. Goldfarb

The tragedy in The Tragedy of the Commons *is perhaps best illustrated by the events surrounding the production of Kepone in Hopewell, Virginia, by the Allied Chemical Corporation and its jobbers. In this statement of the facts surrounding the case, it becomes clear that both Allied and its contractors considered the advantages of "one more cow on the commons" and did not care what the effect on the environment (or in this case even on humans who worked for them) would be.*

HOPEWELL (population approximately 24,000) is an industrial city located on the banks of the James River in southern Virginia. Calling itself "the chemical capital of the South," Hopewell has actively solicited the large chemical concerns. Consequently, Firestone, Hercules, Continental Can, and Allied Chemical have located chemical plants in Hopewell.

In 1949, the initial batch of 500 pounds of a pesticide named Kepone was produced by Allied, and two patents for the process were awarded to it in 1952. Allied did not consider Kepone to be a major pesticide. In fact, Kepone production never exceeded 0.1 percent of America's total pesticide production, and its sales were less than $200,000 annually over a sixteen year period. Kepone was intended primarily for export to Europe for use against the Colorado Potato Beetle, and to South America to control the Banana Root Borer.

Excerpted from Goldfarb, W. *Kepone: A Case Study*, Institute of Government, Department of Political Science, Eastern Kentucky University, 1978.

Before moving to commercial production, Allied inaugurated an extensive series of toxicity tests involving Kepone. Such tests were necessary in order to obtain registration under the federal pesticide laws, which Allied actually received in 1957. The tests were conducted by various consultants, including the Entomology Department of the College of Agriculture at Rutgers University (now Cook College). The results of this research revealed Kepone to be highly toxic to all species tested: it caused cancer, liver damage, reproductive system failure, and inhibition of growth and muscular coordination in fish, mammals, and birds. Upon being presented with the test results, Allied voluntarily withdrew its petition to the Food and Drug Administration for the setting of Kepone residue tolerances for agricultural products.

Kepone is a chlorinated hydrocarbon pesticide, a chemical relative of DDT, Aldrin/Dieldrin, and Mirex (all of which have been banned by the United States Environmental Protection Agency — EPA). As such, Kepone is a contact poison, capable of being absorbed through the skin or cuticle, it is lipophilic (fat soluble), but insoluble in water; it is persistent in the environment; and it will bioaccumulate in the fatty tissues of the body. The exact mechanism by which chlorinated hydrocarbons kill target pests is uncertain. What is known is that they are nerve poisons, interfering with the transmission of electrical impulses along nerve channels. The results of contact with Kepone are loss of control over muscular coordination, convulsions ("DDT-like tremors"), and eventually death.

Despite the unfavorable toxicity test results, Allied deemed Kepone ready for commercial production, and contracted with the Nease Chemical Company of State College, Pennsylvania, to produce it for Allied. The relationship between Allied and Nease lasted from 1958 through 1960. Allied entered into a similar arrangement with Hooker Chemical during the early '60s.

By 1966 even more negative test results had been associated with Kepone, but Allied nevertheless decided to manufacture Kepone on an increased basis in its own Semi-Works facility in Hopewell. In preparation for production an area supervisor of the Semi-Works was asked to develop a production manual. The production manual was to contain operating and safety instructions for the production process. The supervisor naturally consulted available toxicity research results, and his recommended precautions reflect the test findings. At Allied, Kepone spills and dust were closely controlled and workers wore safety glasses and rubber boots and gloves. Allied's Kepone operations were directed by William Moore until 1968, and thereafter by Virgil Hundtofte.

Prior to preparation of the production manual, there had been no recorded case of human exposure to Kepone to the level of acute poisoning. Allied apparently discounted such a possibility, regardless of the documented adverse effects of Kepone on animals. However, a witness for the United States at the trial testified that Allied should have suspected "that the same symptomology would be induced in man if exposed to Kepone."

In 1970, the federal government resurrected the Refuse Act Permit Program, which required all industrial discharges into navigable waters to obtain permits from the U. S. Army Corps of Engineers. The Allied complex at Hopewell had three discharge pipes directly into the stream called Gravelly Run, a tributary of the James River. One of these pipes originated at the Semi-Works where Kepone was manufactured. The Refuse Act Permit (RAP) application was discussed by Allied's plant managers and their assistants, who found themselves on the horns of a dilemma. Allied was discharging Kepone process wastes without treatment of any kind, and the installation of pollution control equipment would be expensive. Moreover, planning was being conducted for the construction of a regional sewage treatment plant which would treat the wastes of all industries in Hopewell, but the municipal treatment plant would not be complete until 1975 at the earliest. What should Allied do during the construction period?

Allied decided to list the Semi-Works discharge as a temporary phenomenon which would be discontinued within two years. In such cases, Federal regulations allowed for a short form RAP application requiring only that the discharge be identified as a "temporary discharge." (Allied gratuitously added that it was unmetered and unsampled.) Thus, neither Kepone nor two plastics products (TAIC and THEIC) also manufactured at the Semi-

Works were listed by Allied on its RAP application, even though Allied quite clearly did not intend to terminate production at Hopewell.

In 1972 the RAP expired, but a new permit program had been enacted — the National Pollutant Discharge Elimination System (NPDES) permit program under the Federal Water Pollution Control Act Amendments of 1972 (FWPCA). The NPDES permit program is administered by EPA instead of the Corps of Engineers.

EPA requested data on the nature, volume, and strength of Allied's discharges, and again the dilemma manifested itself. One of Hundtofte's assistants prepared an option memorandum, outlining three strategies which Allied might follow: 1) to do nothing and hope for a lack of enforcement by EPA; 2) to divert the Semi-Works effluent to another outfall pipe for which a permit had been obtained; or 3) to slowly improve the Semi-Works effluent so as to "buy time" until completion of the municipal system. None of these options, however, entailed the disclosure by Allied of its Semi-Works effluent. The last of these options was selected, and again Allied submitted data to the federal government describing the Semi-Works discharges as unmetered, unsampled and temporary outfalls. As a result, between 1966 and 1974 Allied discharged untreated Kepone and plastics wastes into Gravelly Run without revealing the nature of its discharges to the federal government.

In 1973 Allied underwent a corporate reorganization, during which control of the Semi-Works facility was transferred from the Agricultural Division to the Plastics Division. The former was superseded in expectation of its impending move to new facilities in Baton Rouge, Louisiana. Virgil Hundtofte, Plant Manager of the Agricultural Division at Hopewell, and William Moore, Research Director, made plans to retire from the company rather than to relocate. (Hundtofte had been with Allied in Hopewell since 1965, Moore since 1948.)

One effect of the reorganization was a reorientation of production priorities among the products manufactured at the Semi-Works. Kepone production had decreased steadily, but THEIC, which had been manufactured in small quantities for eighteen years, suddenly found a lucrative market calling for a doubling of production. THEIC and Kepone shared certain production equipment, and with the surge in demand for THEIC a decision was made in 1973 to "toll" Kepone production. Tolling is a common arrangement in the chemical industry whereby another company performs processing work for a fee or a "toll" and then returns the final product. The keynote of a tolling arrangement is that during the processing period legal title to the materials and product remains with the supplier, in this case Allied.

In January, 1973, when the decision to toll Kepone was divulged, William Moore saw his opportunity to remain in Virginia and continue in the Kepone manufacturing business. He immediately contacted Hundtofte, who had recently resigned from Allied and gone to work for a fuel oil distributor. Moore and Hundtofte agreed to form a corporation and bid for the Kepone tolling contract. On November 9, 1973, Life Science Products Company (LSP) was incorporated under the laws of the Commonwealth of Virginia. Moore and Hundtofte were the only shareholders, directors and officers of LSP. Less than a month later, the tolling agreement between Allied and LSP was signed. Allied had solicited bids from Hooker Chemical, Nease Chemical, Velsicol, and LSP, but LSP's bid was by far the lowest: 54 cents per pound for 500,000 pounds of Kepone. Nease Chemical (which, it may be recalled, manufactured Kepone for Allied from 1958 through 1960) declined to bid, but responded that if it chose to bid on the contract it would cost Nease 30 cents per pound for waste disposal alone. Hooker (Nease's successor) bid $3.00.

The details of the tolling agreement are important because the question of Allied's responsibility for LSP's illegal acts loomed large at the trial. The contract provided that Allied would supply — at its own expense — all of the raw materials for Kepone production, with the title to remain in Allied. Within certain broad limits, Allied would determine the monthly production rate of Kepone, which would be packed in Allied containers and transported in Allied trucks. Allied also agreed to pay LSP's taxes, other than corporate income taxes. LSP was to receive between 32 and 38 cents per pound for 650,000 pounds or more of Kepone.

Through a capital surcharge arrangement, Allied was to pay for all of LSP's approved capital expenditures, whether for production or pollution control, except for land and building. If LSP closed for pollution violations during the first year of the contract, Allied had the option to purchase LSP's assets for $25,000. And if the contract was terminated by either party for any reason, LSP agreed to refrain from producing Kepone for anyone else.

The relationship between Allied and LSP was only partially defined by the tolling agreement. Moore and Hundtofte promised Allied that they would not dispose of their shares in LSP without Allied's consent. Moreover, Allied assisted LSP in many ways: to obtain loans and equipment from outside sources; to meet temporary cash deficits; to augment fuel supplies during the oil embargo; to attain greater efficiency by the use of Allied facilities. Most importantly, LSP's effluent was sampled and analyzed by Allied personnel after Virginia ordered such testing. Before that (up to October, 1974), LSP had tested its effluent by a visual check — if the effluent was cloudy, the presence of suspended Kepone was indicated.

Allied officials regularly toured the LSP plant, and were informed by mail of the waste disposal problems which LSP faced almost from its inception. Whereas Allied had discharged the residues of its Kepone production process directly into tributaries of the James River, LSP decided to discharge into the Hopewell sewer system, despite the fact that the regional treatment plant was still under construction. By this means, LSP could avoid having to apply for an NPDES permit, which is not required of "indirect discharges" (discharges into municipal treatment systems). LSP contacted C.L. Jones, Director of Hopewell's Department of Public Works, for permission. At that time, Hopewell possessed a primary waste treatment plant — a series of settling tanks without biological or chemical treatment other than disinfection and sludge digestion. Such a rudimentary system could not degrade Kepone, but would merely divide Kepone influent between outfall pipe and sludge. Jones, who had been Plant Manager of Allied's Hopewell Semi-Works prior to Hundtofte, recommended to Hopewell's City Manager that Hopewell accept LSP's wastes. (In order to assure that no damage would be done to the sludge digester at the treatment plant, LSP was asked by Hopewell to meet a pretreatment standard of three parts per million of Kepone.) Thus, LSP became the only industry in Hopewell allowed to discharge into the municipal sewerage system.

LSP commenced operation in March, 1974, and almost immediately, large quantities of Kepone began flowing into Hopewell's treatment plant. In October, a state inspector discovered that the sludge digester at the plant was inoperative, and his investigation revealed LSP to be the source of contamination. Prior to the plant breakdown, the state was apparently unaware that Kepone was being discharged into the Hopewell system because Hopewell's application for an NPDES permit for its treatment plant (filed on October 10, 1973 — about a month before plug-in permission was granted to LSP) made no mention of any industrial discharge to the municipal system. Moreover, additional information filed by Hopewell in July, 1974, did not include notice of LSP's discharges. When the State brought the situation to the attention of LSP and Hopewell officials, LSP's discharges were not prohibited even though the pretreatment standard was being violated, but a study was commenced to determine a "safe" effluent limit for Kepone. Finally, in June of 1975, a more restrictive pretreatment standard was imposed upon LSP (0.5 parts per billion). EPA, which had been informed of the situation, agreed to this compromise, even though it had earlier recommended a stricter effluent limitation. In order to meet this standard LSP had to further pretreat its wastes and hold its discharges in tanks until such time as the discharge would not violate the pretreatment standards (i.e. even out the flow). Allied had participated in the negotiations among LSP, Hopewell, Virginia, and EPA; and Allied opted to pay for the necessary pollution control equipment. Allied and LSP then began to discuss the capital costs of expanding Kepone production to 2,500,000 pounds per day in order to meet an increasing demand in the European market. (From the inception of LSP, Allied had constantly requested increased Kepone production.) However, even after the new equipment was installed, the pretreatment standard was violated in nineteen out of twenty-one samplings.

On July 7, 1975 as LSP was preparing for increased Kepone production, one of its employees visited an Hopewell internist named Chou, complaining of tremors, weight loss, quickened pulse rate, unusual eye movements, and a tender, enlarged liver. Such symptoms were not unusual among LSP employees, but were generally dismissed as "the shakes" — a necessary price to be paid for the $5.00 per hour wage they received. Although about twenty physicians had been consulted during the sixteen months of LSP's existence, only Dr. Chou suspected a connection between the ailments and the workplace environment. After questioning his patient and taking a blood sample, Dr. Chou forwarded the sample to the Center for Disease Control in Atlanta, where he knew that an analysis for Kepone could be performed. The tests disclosed that the blood sample contained 7.5 parts per million of Kepone, an astounding concentration to be found in human blood. Federal doctors then contacted the Virginia State Epidemiologist, Dr. Robert Jackson, who quickly arranged a meeting with Hundtofte and Moore. Similar meetings had previously been requested by the Virginia State Department of Labor and Industry, but LSP had been successful in postponing them. Even though a former LSP worker had filed a complaint with the United States Department of Labor, that agency had also failed to inspect the LSP factory. Only one Federal official, an EPA pesticide inspector, visited LSP before July, 1975; but he was not authorized to enter the production area. Other media under EPA jurisdiction — air and water pollution — had been delegated to state and local officials. At various times representatives of the Virginia Air Pollution Control Board, Water Pollution Control Board, and State Health Department had visited LSP; but they were not responsible for inspecting LSP's working conditions.

When Jackson toured the plant he was appalled: "Kepone was everywhere;" conditions were "incredible;" workers were virtually "swimming in the stuff." Workers wore no protective equipment, nor had warning signs been posted. Seven out of ten production workers present had "the shakes" so severely that they required immediate hospitalization. On July 25, 1975, LSP voluntarily closed out its operation under threat of a closure order by the

Virginia Department of Health. Further investigation divulged seventy-five cases of acute Kepone poisoning among LSP workers and high levels of Kepone in the blood of some of their family members. Moreover, there was found to be massive contamination of air, soil, and especially water in the vicinity of the plant. As a result. the State of Virginia closed one hundred miles of the James River and its tributaries to fishing.

(*Ed. Note:* Allied was eventually convicted on over 1000 counts of pollution, and paid a substantial fine. Nobody went to jail and nobody lost their job. In the next annual report following the payment of the fine, Allied reported that this wasn't a big thing and they could easily cover the fine from operating revenue.)

2.4 The Hooker Memos

The following is a direct transcript from the CBS program 60 Minutes. Hooker Chemical Company, a subsidiary of Occidental Petroleum, appeared to have for years been disposing of toxic materials into the ground and polluting the groundwater. Such contamination is considerably more serious than the pollution of surface waters (as was the case in Hopewell) since the contamination of groundwater is for all intents and purposes permanent. What is interesting in this article is the apparent ignorance of the company executives as to what was really taking place. One almost feels sympathy for Mr. Baeder, the president of Hooker. The fault seems not to be so much his as in the inability of middle managers to act on serious but clearly defined problems which may not have been in the company operations manual. Management was not impressed by appeals for reason, even when written in terms of potential financial liability to the company. For middle managers to heed the concerns expressed in the memos from Mr. Edson would have required a special appreciation of and concern for the environment — an enlightened environmental ethic.

MIKE WALLACE: Tonight, ''The Hooker Memos'' provide a rare inside look into an American company that some have called a ''corporate outlaw.'' The company is the giant Hooker Chemical. Even before these memos surfaced, Hooker's image was in trouble. Back in 1977, residents of Love Canal in Niagara Falls had to evacuate their homes because soil contaminated by Hooker's hazardous waste had seeped into their basements. Last October,

Reprinted from the 16 December 1979 broadcast of *60 Minutes*, CBS News, New York. Used with permission.

Hooker said it would spend $15 million to clean up damage caused by its chemical pollution of a lake in the State of Michigan. And now the State of California is investigating Hooker on charges that the company's Occidental Chemical Plant at Lathrop, California, for years dumped toxic pesticide waste in violation of state law, polluting the ground water nearby. Hooker officials deny the charges, but Hooker memos seem to say the company knew what it was doing and, nonetheless, just kept on doing it.

The man who wrote these memos, Robert Edson, is Hooker's environmental engineer at their Lathrop plant. He wouldn't speak to us, but his memos speak for themselves.

On April the 29th, 1975, Robert Edson wrote: "Our laboratory records indicate that we are slowly contaminating all wells in our area, and two of our own wells are contaminated to the point of being toxic to animals or humans. THIS IS A TIME BOMB THAT WE MUST DE-FUSE."

June 25th, 1976, a year later, there was this: "To date, we have been charging waste water … containing about five tons of pesticide per year to the ground … I believe that we have fooled around long enough and already overpressed our luck."

A year after that, on April 5th, 1977, Edson was still writing memos. "The attached well data" he says, "shows that we have destroyed the usability of several wells in our area. If anyone should complain, we could be the party named in an action by the Water Quality Board. … Do we correct the situation before we have a problem or do we hold off until action is taken against us?"

And on September 19th, 1978, more than a year later, Edson again pointed out to his management: "We are continuously contaminating the ground water around our plant."

And this is that plant, the huge Occidental Chemical complex where Bob Edson's memos say chemical waste was being dumped and was poisoning the ground water. Back in 1968, the Water Quality Control Board for the region came to an understanding with Oxy Chem about what this plant would be permitted to discharge as waste. Oxy Chem agreed to discharge no substance that would cause harm to human, plant or animal life. But back then, the water board thought that Oxy Chem was dealing only with fertilizers and nontoxic fertilizer waste. It turned out that was not the case. What the water board did not know, had no idea of, was that another kind of waste was being dumped in here, toxic waste from pesticides, which were also manufactured over at that plant but were nowhere mentioned in Occidental's agreement with the water board. Hooker Chemical, itself a subsidiary of the giant Occidental Petroleum, headquarters in Houston, Texas, and we wondered if Hooker's president Don Baeder, would talk to us about the Lathrop Plant and those Edson memos. He would. But he maintained that the water board in California had known about and had okayed Occidental Chemical's discharge of pesticide wastes. And he told us that Robert Edson simply didn't know that.

DON BAEDER: Mike, it's unfortunate, but Mr. Edson didn't take on the environmental assignment until 1972, so he was unaware of this visit by the state to our facilities. And ...

WALLACE: May I see that a second?

BAEDER: Yes, you may. I wish you'd read that last paragraph, too.

WALLACE: "We wish to thank Occidental Chemical for the spirit of concern and cooperation in which this problem was met and to commend you for the thoroughness of your approach." Signed Charles Carnahan, the executive officer of the California Regional Water Quality Control Board. And this is dated 4 September 1970.

BAEDER: Does that sound like a company that was not concerned with the environmental problems back in 1970, Mike?

WALLACE: I'm at a loss, then, to understand why your own Mr. Edson would say some five years later, "Recently published California State Water Quality Control laws state we cannot percolate chemicals to ground water. The laws are extremely stringent about pesticides. And to date, the Water Quality Control people don't know about our pesticide waste percolation." This is your own man.

BAEDER: Yes, and Mr. Edson ...

WALLACE: Chief environmental engineer for Oxy Chem at Lathrop.

BAEDER: Mr ... and Mr. Edson is a good man. And Mr. Edson somehow felt it necessary to ... to steel the management to action with these kind of statements. But they didn't need steeling.

WALLACE: June 25th, 1976, a memo from R. Edson to A. Osborne. Who's A. Osborne?

BAEDER: I believe he's another engineer at the plant.

WALLACE: "For years," he says, "we've dumped waste water containing pesticides and other ag-chem products to a pond southwest of our plant. Our closest neighbor's drinking water well is located less than 500 feet from the subject percolation pond and, fortunately for the management, no pesticide has yet been detected in his water. I personally would not drink from his well." This is Edson, this fellow that you respect so.

BAEDER: I do.

WALLACE: This fellow who should have known what ... Why didn't you show him this letter that you've shown me?

BAEDER: I don't ... I think the important thing, Mike, is ...

WALLACE: But ...

BAEDER: ... what actions that we took to ...

WALLACE: No, wait just a second. Why didn't you, someplace

along the line, say, "Hey, Bob. Don't you know that the California Water Quality Control Board back in 1970 sent us this letter? They — you're ... you're ... you're talking through your hat, Bob Edson, because the California people knew all about this all the time.

BAEDER: The fact that certain of our people in our organization didn't know about it isn't ... isn't germane. The fact is the Water Quality Control Board did know about it.

WALLACE: But Charles Carnahan, the man who signed that letter Don Baeder has shown us, Charles Carnahan, the executive officer of the water board, told us that when he wrote that letter, he had no idea that Occidental was dumping toxic pesticide waste.

CHARLES CARNAHAN: The waste that they were dumping we figured was the waste from their fertilizer manufacturing.

HARRY MOSES: How do you feel about it now that you've learned differently?

CARNAHAN: Well. I feel kind of stupid. I feel stupid because I think that we took them at ... on good faith and we were fooled.

WALLACE: But Don Baeder, who still maintained that the water board did know, also insisted that no harm was done, and that, he said, is what is really important.

BAEDER: No one has been hurt at Lathrop. No one. From our discharges, no one has been hurt and no one will be. No one will be.

DR. ROBERT HARRIS: When you contaminate ground water, you do so, for all practical purposes, irreversibly.

WALLACE: Dr. Robert Harris is an expert on toxic chemicals, recently appointed to the President's Council on Environmental Quality.

DR. HARRIS: The reason is that ground water moves very slowly, and that these contaminants that are now present in the ground water will continue to be there, will continue to migrate towards population centers, toward individual wells, for decades to come. This water will not be cleansed easily.

WALLACE: This past summer, the folks here at Oxy Chem drilled 12 test wells to find out just how contaminated the ground water had become. Five of the wells showed quantities of a pesticide called DBCP, and this well, number seven well, contained it in excessive amounts. Enough, according to Dr. Harris, to increase the risk of dying from cancer by as much as 25 percent if the water from this well were consumed over a normal lifetime.

DR. HARRIS: These are not small concentrations of these particular chemicals.

WALLACE: Is it possible they just didn't know, that the state of the art at the time that this kind of thing went on was such that they didn't know what they were doing?

DR. HARRIS: They should have known. They should have known that DBCP had and did cause sterility in laboratory animals more than a decade ago, and they should have known that DBCP was being tested at the National Cancer Institute, and that preliminary results showed that it was a very potent carcinogen as early as 1973. This is information that was available to the general public, and I assume was ... was made available to Hooker.

BAEDER: As soon as we found out it caused sterility, as soon as we found out it caused cancer, those operations were shut down.

WALLACE: When was that?

BAEDER: That was in 1977, I believe, the summer of '77.

WALLACE: Shortly after you took over as president?

BAEDER: Yes.

WALLACE: All right. And you don't want to go back into the business of making DBCP?

BAEDER: We are not going back into the business of making DBCP.

WALLACE: You don't want to go back in?

BAEDER: I don't want to.

WALLACE: Well, then, how come I have a memo here, dated December 11th, 1978, to D.A. Guthrie from J. Wilkenfeld, subject: "Re-entry to DBCP Market"? Why would you ... why would your company inquire into the possibility of — quote — "re-entry into the DBCP market" if you don't want it? You're the president. This happens in 1978.

BAEDER: Mike ...

WALLACE: Why ... why would they be exploring it, Mr. Baeder?

BAEDER: Mike, if the government permits it and we develop safe ... safe systems for handling it to eliminate any exposure, any potential harm to people, it ... it could be and will be produced. It's an important ... it's an important chemical to agriculture of California. Now just because if something is mishandled it is toxic or it causes cancer does not mean that that same material cannot be handled safely. And I can assure you that the State of California would not allow anyone to produce that under risk of these kinds of

... of ... of problems without markedly changing the procedures under which it was produced.

WALLACE: It is my understanding, according to this memo here, that the State of California has rejected requests for permission to use the material, and the government of Mexico has shut down the two manufacturing plants for DBCP that were operating in that country. And yet, your Acting Vice President for Environmental Health and Safety, Mr. David Guthrie, says "We've reviewed the proposal for re-entry into the DBCP business and we have no environmental health objections. Jerry Wilkenfeld has no technical objections." Who's Wilkenfeld?

BAEDER: Mr. Wilkenfeld is a ... is responsible for environmental health safety at the Occidental level.

WALLACE: For environmental health and safety?

BAEDER: At the Occidental ...

WALLACE: Let me read you what he says.

BAEDER: You did read it.

WALLACE: "Assume..." Oh, no, I haven't read the whole thing.

BAEDER: Oh.

WALLACE: "Assume that 50 percent of the normal rate for these people exposed may file claims of effects from the exposure. Determine the number of potential claims for sterility and cancer, based on the insurance department's or legal department's estimate of the probable average judgment or settlement which would result from such a claim. Calculate the potential liability, including 50 percent for legal fees and other concise ... contingencies." One gets the impression that profits are more important to the Hooker Chemical Company than care for human health.

BAEDER: We went out of DBCP as soon as we found out it presented any harm or exposures to the people or the workers.

WALLACE: And a year ago you were back in the business of looking at the possibility of going back into the business.

BAEDER: People are looking at it, but we are not into it. And that would have been a very deliberate corporate decision to go back into it. And I ... look, Mike, before I'm a president, I'm a human being.

WALLACE: Uh-hmm.

BAEDER: We would not have gone back into it. Again, you're — you're dealing in studies that people make. We make a lot of studies. Why do you hammer us on something that might have happened but hasn't happened?

WALLACE: The only reason I'm hammering it is: Is this the way that America does business?

BAEDER: America looks at many options in doing business. It looks at many options.

WALLACE: And one of the options is, can we afford ...

BAEDER: No. No. Mike, there is a risk in making almost an ... there's a risk in making drugs. Mike, your drug people look at this same thing in every pharmaceutical drug they put on the market. There is a risk. There ...

WALLACE: Do you know how this ... this memo ends up? It says, "Should this product" — DBCP — "still show an adequate profit, meeting corporate investment criteria, then the product should be considered further." That's the bottom line.

BAEDER: Mike, I tried to tell you that profit is not the primary consideration. Mike ...

WALLACE: It's your own memo.

BAEDER: Mike, it's a memo by a ... a young man in the corporation. It's not the policy of the corporation. Young people do not set policy. Mike, I tried to tell you that over the last three years in environmental health and safety, we've spent more than a hundred and thirty million dollars. That's more than our profits have been.

WALLACE: I wonder why.

BAEDER: Why? Because ...

WALLACE: Yeah. Why have you had to?

BAEDER: ... because environment is important. Mike, we're a ... you know, we ... we have a concern for our people.

WALLACE: Fine ...

BAEDER: ... we have a concern for our neighbors, and that's why we're spending this money.

REP. ALBERT GORE (D-Tennessee): Well, I want to believe him, and I hope that what he says is true. I hope that this company and the chemical industry generally will accept the very great public responsibility that it has.

WALLACE: Congressman Albert Gore is a member of the House subcommittee that has been holding hearings on hazardous waste disposal, and he is genuinely alarmed by the extent of the problem.

REP. GORE: Two or threes decades from now, in many parts of the country we'll be facing widespread shortages. This is a very precious resource which must be conserved and protected, yet we are systematically poisoning it by dumping 80 billion pounds every year into the ground. It was formerly thought that the ground was just like a big sponge that could soak up all the poison that we could pour into it, but it's not the case.

WALLACE: How bad is Hooker Chemical, Congressman? Is it one of the worst or just in the middle of the catalogue of chemical corporate outlaws?

REP. GORE: The ... the problem is industrywide. They are every day making hard calculations, just like this company did, as to whether or not the risk to other Americans is worth it for them to make a good deal more money.

BAEDER: The State of California, most states, permit certain discharges to the environment, because it's their belief that they can be tolerated. When we learn otherwise, we change. The problem we have is that, with today's knowledge, there's no question we would have done something differently. The real issue is: Should we be judged by today's knowledge on past practices? And I think that's the issue. I think the American people have got to understand that.

GEORGE DEUKMEJIAN: The evidence appears to us to indicate very clearly that they have done this willfully and knowingly.

WALLACE: George Deukmejian, California's attorney general, is currently asking that Hooker Chemical pay millions of dollars in fines. But are fines an effective deterrent? Or, we wondered, would jail terms for corporate officers whose companies break the law be a more effective deterrent?

DEUKMEJIAN: I don't really feel that we need to have additional laws as much as we need to enforce the laws that we do have. And I think that if we do that, I feel that we will be able to control this type of practice. Also, when you get into the criminal law area, you're talking about the need, I suppose, to prove a criminal intent, and that might be difficult, if not im ... impossible, to do in a case of this type.

WALLACE: What do you think about the suggestion of criminal penalties for corporate outlaws?

BAEDER: Mike, I think that criminal penalties for criminal acts are completely justified.

WALLACE: Is it a criminal act to poison, sterilize, whatever, if it can be proved that it was done with knowledge?

BAEDER: Mike, if it can be proved that causing sterilization by violating the law, by ... by operating in a way that you know is unsafe ...

WALLACE: Hm-mmm?

BAEDER: ... I would say that's a criminal act.

WALLACE: And who should serve the time or pay the fine?

BAEDER: I think the people that are involved?

WALLACE: The top man or ...

BAEDER: I think the buck always stops at the top, sir.

WALLACE: Since we filmed this report, there has been this late development. California Attorney General George Deukmejian will file a lawsuit this coming week against Hooker Chemical. The suit will ask Hooker to pay fines and clean-up costs in excess of $15 million for the damage caused by their Lathrop plant.

(*Ed. Note:* The Occidental Petroleum Company is presently in the process of spending millions of dollars to extract contaminated water from the ground so that the plume of pollution will not extend any farther.)

2.5 The Bunker Hill Lead Smelter

C. Tate

In the case of Hooker Chemical management, conscious decisions were made to ignore the warnings of impending environmental disaster. It was not a case of not knowing. It was a case of deliberately choosing to act in what we would now consider an environmentally destructive manner. But what if there were true conflicts where several values of equal importance conflicted? What if the problems were not of the nature of environment vs. profits?

In the situation described below, just such a dilemma is faced by the management of a primary materials industry. What is not stated in the article is the nature of the business and how precariously such industries remain operational in the United States. Not only is there a question of job safety, but a question of employment opportunity in the community. And in some cases, even the survival of the community. Dilemmas such as described below are daily fare for engineers who have attained management positions.

LEAD IS ONE of the oldest, most ubiquitous, most toxic, and most thoroughly studied substances used by man. Lead can damage the nervous system, the kidneys, and the reproductive organs. At sufficiently high levels, it causes convulsions, coma, and death. The effects of lead begin at low levels, so that although there may

---

Excerpted from Tate, C. "The American Dilemma of Jobs, Health in an Idaho Town," *Smithsonian*, September, 1981. Used with permission.

be no observable symptoms, damage may have been done, and it may be irreversible.

* * *

"There's nothing wrong with my kids," says Kathy Kriedeman flatly. Kriedeman, her husband, Lowell, and their two children live in a small, tidy house on the main street of Smelterville, a community with a post office, two taverns, about 840 citizens, and some of the highest concentrations of lead in the Kellogg area. Her children, now aged 9 and 13, both had lead levels higher than 70 micrograms when tested during the Center for Disease Control (CDC) survey. Kriedeman — a large, sturdy woman in her early thirties, with mild blue eyes — has refused to have them participate in any of the numerous follow-up surveys since then. She has declined several offers to have them tested for neurological or psychological abnormalities. "I don't like all these people poking at my kids, sticking their noses in where they don't belong."

Kathy stayed home until her children were out of the toddler stage. Then she joined her husband at "the Bunker," getting a job in the smelter.

The smelter had not hired women — except for a brief period during World War II — until the early 1970s when it was opened up in response to pressure from the Equal Employment Opportunity Commission. Kriedeman was one of the first to be hired. The Occupational Safety and Health Administration (OSHA), the federal agency that sets and enforces standards for workplace health and safety, has measured lead levels up to 7.37 milligrams — 7,370 micrograms — per cubic meter of air over an eight-hour period.

Respirators were available to smelter employees at the time, but their use was not strictly enforced. Kriedeman rarely wore hers. "They're too hard to breathe through, especially when you're working." Ten months after beginning her job, she discovered she was pregnant and asked for a transfer. "There was all that smoke coming directly up in my face. I was getting sick from it." But, she says, her request was denied, so she quit her job. A week later, she miscarried. She says her doctor, since deceased, told her that the miscarriage had probably occurred because she had been working in the smelter.

Kriedeman returned to work several months later. She worked until April 1975, when Bunker Hill announced that women capable of bearing children would no longer be employed in he smelter because of "medical information which indicated lead might cause fetal complications." This meant that a woman of child-bearing capability would have to undergo sterilization in order to be eligible to work in the smelter. The 29 women then at work were transferred to other jobs. Eighteen of them, including Kriedeman, promptly filed charges of sex discrimination, to which an exasperated James H. Halley, Bunker Hill's president at the time, responded:

"So, am I supposed to put women in the plant, or am I supposed to keep them out? Which is the most moral thing to do? If we don't put women in the smelter that's going to mean fewer jobs for women. If we put women in the smelter, and she gets pregnant, we're liable to have a mentally retarded person born who otherwise would have been normal."

2.6 The Existential Pleasures of Engineering
Samuel Florman

*The anti-technology movement in the 1960s was aided in great part by the inability of engineers to articulate what they did for a living and why. To many, the ills of our society were **caused** by the engineers, who were responsible for such items as nuclear bombs, pesticides, and airplane crashes. Samuel Florman, an eminent civil engineer, rose to the defense of engineering in his widely-read book* The Existential Pleasures of Engineering. *He points out first that engineering is **fun**. "What a startling word." He continues, "Engineering is **fun**, and similar to the creative arts in providing fulfillment." In the section of his book reproduced below, he argues for this aspect of engineering, and then goes on to show that the nature of engineering has been misconceived. "Analysis, rationality, materialism, and practical creativity do not preclude emotional fulfillment; they are pathways to such fulfillment. Engineering is superficial only to those who view it superficially. At the heart of engineering lies existential joy."*

*Perhaps many engineers will agree that engineering is indeed fun. But what is this **existential** business? Isn't this totally alien to the spirit of engineering? Not so, insists Florman, as long as one interprets existentialism as the search for inner truth, the elimination of external influence, and the disenchantment with conventional creed and delusions. The engineer, in his/her search for the best solution to technical problems is, according to Florman, free of such externalities, and thus has existential freedom to perform his/her task.*

Excerpted from Florman, S. *The Existential Pleasures of Engineering*, St. Martin's Press, New York, 1976. Used with permission.

*This all sounds attractive, until the question of social and environmental responsibility is raised. Florman's philosophy, in our opinion does not deal adequately with such questions. As perhaps the most gruesome example to illustrate this problem, consider the fact that the gas ovens in Nazi extermination camps were designed by **engineers**. They were simply given a job to do, and they went about doing it, free of any social concern. The construction of unneeded, uneconomical and environmentally disastrous dams, waterways, and other facilities by the Corps of Engineers differs both in scale and substance, but not in philosophy. Engineers have fun — yes — but often at the expense of other humans or the environment.*

THE FIRST and most obvious existential gratification felt by the engineer stems from his desire to change the world he sees before him. This impulse is not contingent upon the need of mankind for any such changes. Doubtless the impulse was born from the need, but it has taken on a life of its own. Man the creator is by his very nature not satisfied to accept the world as it is. He is driven to change it, to make of it something different. Paul Valery, in his poetic drama, *Eupalinos*, has expressed this impulse with a romantic flourish:

> The Constructor ... finds before him as his chaos and as primitive matter, precisely that world-order which the Demiurge wrung from the disorder of the beginning. Nature is formed and the elements are separated; but something enjoins him to consider this work as unfinished, and as requiring to be rehandled and set in motion again for the more special satisfaction of man. He takes as the starting point of his act the very point where the god left off ... the masses of marble should not remain lifeless within the earth constituting a solid night, nor the cedars and cypresses rest content to come to their end by flame or by rot, when they can be changed into fragrant beams and dazzling furniture.

This desire to change the world is brought to a fever pitch by the inertness of the world as it appears to us, by the very *resistance* of inanimate things, to use the concept expressed by Gaston Bachelelard in *La Terre et les Reveries de la Volonte*:

> The resistant world takes us out of static being.... And the mysteries of energy begin.... The hammer or the trowel in hand, we are no longer alone, we have an adversary, we have something to do.... All these *resistant* objects ... give us a pretext for mastery and for our energy.

The existential impulse to change the world stirs deep within the engineer. But it is a vague impulse that requires particular projects for its expression. Here the engineer cannot help but be enthralled by the countless possibilities for actions that the world presents to him. In *A Family of Engineers*, Robert Louis Stevenson has told of the allure that the profession of engineering had for his grandfather:

> ... the perpetual need for fresh ingredients stimulated his ingenuity.... The seas into which his labors carried the new engineer were still scarce charted, the coast still dark.... The joy of my grandfather for his career was as the love of a woman.

The engineer today, for all his knowledge and accomplishment, can still look out on seas scarce charted and on coasts still dark. Each new achievement discloses new problems and new possibilities. The allure of these endless vistas bewitches the engineer of every era.

For many engineers, the poetic image of seas and coasts can be taken literally. Water and earth are the substances that engaged the energies of the first engineer — the civil engineer. Civil engineering is the main trunk from which all branches of the profession have sprung. Even in this age of electronics and cybernetics, approximately 16 percent of American engineers are civil engineers. If we add mining, basic metals, and land and sea transportation, fully a quarter of our engineers are engaged in the ancient task of grappling with water and earth. Civil engineering has traditionally included the design and construction of buildings, dams, bridges, railroads, canals, highways, tunnels — in short, all engineered structures — and also the disciplines of hydraulics and sanitation: water supply, flood control, sewage disposal, and so forth. The word "civil" was first used around 1750 by the British engineer, John Smeaton, who wished to distinguish his works (most notably the Eddystone Lighthouse) from those with military purposes. The civil engineer, with his hands literally in the soil, is existentially wedded to the earth, more so than any other man except perhaps the farmer. The civil engineer hero of James A. Michener's novel, *Caravans*, cries out, "I want to stir the earth, fundamentally ... in the bowels." The hydraulic engineer hero of Dutch novelist A. Den Doolaard's book, *Roll Back the Sea*, stares across the flood water rushing through broken dikes and feels "a strange and bitter joy. This was living water again, which had to be fought against."

Living water. Nature, which appears at one moment to be inert and resistant, something which the engineer is impelled to modify and embellish, in the next instant springs alive as a flood, a landslide, a fire, or an earthquake, becomes a force with which the engineers must reckon. Beyond emergencies and disasters, through the environmental crisis of recent years, nature had demonstrated that she is indeed a living organism not to be tampered with unthinkingly. Nature's apparent passivity, like the repose of a languid mistress, obscures a mysterious and provocative energy. The engineer's new knowledge of nature's complexities is at once humbling and alluring.

Another dichotomy with which nature confronts the engineer relates to size. When man considers his place in the natural world, his first reaction is one of awe. He is so small, while the mountains, valleys and oceans are so immense. He is intimidated. But at the same instant he is inspired. The grand scale of the world invites him to conceive colossal works. In pursuing such works, he has often shown a lack of aesthetic sensibility. He has been vain, building useless pyramids, and foolish, building dams that do more harm than good. But the existential impulse to create enormous structures remains, even after he has been chastened. Skyscrapers, bridges, dams, aqueducts, tunnels — these mammoth undertakings appeal to a human passion that appears to be inextinguishable. Jean-Jaques Rousseau, the quintessential lover of nature undefined, found himself under the spell of this passion when he came upon an enormous Roman aqueduct:

> I walked along the three stages of this superb construction, with a respect that made me almost shrink from treading on it. The echo of my footsteps under the immense arches made me think I could hear the strong voices of the men who had built it. I felt lost like an insect in the immensity of the work. I felt, along with the sense of my own littleness, something nevertheless which seemed to elevate my soul; I said to myself with a sigh: "Oh! that I had been born a Roman!" ... I remained several hours in this rapture of contemplation. I came away from it in a kind of dream....

The rapture of Rousseau for the "immensity of the work" survives in the midst of our most bitter disappointment with technology. A 1964 photo exhibition at The Museum of Modern Art in New York, entitled "Twentieth-Century Engineering," brought home this truth to a multitude of viewers. The introduction to the exhibition catalogue directed attention to the fact that the impact of enormous engineering works is sometimes enhanced by the "elegance, lightness, and the apparent ease with which difficulties are overcome," and sometimes by the opposite, the monumental extravagance that appears when "the engineer may glory in the sheer effort his work involves." Ada Louise Hustable of the *New York Times* reacted to the show with an enthusiasm that even the proudest of civil engineers would hesitate to express:

> It is clear that in the whole range of our complex culture, with its self-conscious aesthetic kicks and esoteric pursuit of meanings, nothing comes off with quite the validity, reality, and necessity of the structural arts.

Other art forms seem pretty piddling next to dams that challenge mountains, roads that leap chasms, and domes that span miles. The kicks here are for real. These structures stand in positive, creative contrast to the willful negativism and transient novelty that have made so much painting and literature, for example, a kind of diminishing, naughty game. The evidence is incontrovertible: building is the great art of our time.

"The kicks here are for real." And if they are for real to the observer of photographs, imagine what they are like to the men who participate in creating the works themselves. *Roll Back the Sea*, the Dutch novel already mentioned, has a scene describing the building of the Zuyder Zee wall which gives some slight taste of the excitement surrounding a massive engineering work:

The great floating cranes, dropping tons of still clay into the splashing water with each swing of the arm. Dozens of tugboats with their white bow waves. Creaking bucket dredges; unwieldy barges; blowers spouting the white mass of sand through long pipeline out behind the dark clay dam; and the hundreds of polder workmen in their high, muddy boots. An atmosphere of drawing boards and tide tables, of megaphones and jingling telephones; of pitching lights in the darkness, of sweat and steam and rust and water, of the slick clay and the wind. A dike in the making, the greatest dike that the world had ever seen built straight through sea water.

The mighty works of the civil engineer sometimes appear to be conquests over a nature that would repel mankind if it could. Thus Waldo Frank perceived the Panama Canal slashing through the tropical jungle:

Its gray sobriety is apart from the luxuriance of nature. Its willfulness is victor over a voluptuary world that will lift no vessels, that would bar all vessels.

At other times the civil engineer's structures appear to grow out of the earth with a natural grace that implies that fulfillment of an organic plan. Pierre Boulle, in *The Bridge Over the River Kwai*, writes: "An observer, blind to elementary detail but keen on general principles, might have regarded the development of the bridge as an uninterrupted process of natural growth." The bridge rose day by day, "majestically registering in all three dimensions the palpable shape of creation at the foot of these wild Siamese mountains..." Fifty years after the construction of the Eiffel Tower a Parisian recalled: "It appeared as if the tower was pushing itself upward by a supernatural force, like a tree growing beyond bounds yet steadily growing.... Astonished Paris saw rising on its own grounds this new shape of a new adventure."

From the organic implications of the civil engineer's structures it is but a short step to the spiritual. Mighty works of concrete, steel, or stone, seeming alive but superhuman in scope, inevitably invoke thoughts of the divine. The ultimate material expressions of religious faith are, of course, the medieval cathedrals. They are usually defined as the material creations of religious men. But they can also be considered as magnificent works of engineering which, through their physical majesty and proportion, impel the viewer to think lofty thoughts. In *Mont Saint-Michel and Chartres* Henry Adams has conveyed a sense of the way in which these physical structures both reflect and evoke a spiritual concept:

Every inch of material, up and down, from crypt to vault, from man to God, from the universe to the atom, had its task, giving support where support was needed, or weight where concentration was best, but always with the condition of showing conspicuously to the eye the great lines which led to unity and the curves which controlled divergence; so that, from the cross on the fleche and the keystone of the vault, down through the ribbed nervures, the columns, the windows, to the foundations of the flying buttresses far beyond the walls, one idea controlled every line.

William Golding, in his novel *The Spire*, has explored the

theme of the interrelationship between construction and religion. Set in medieval England, the novel relates the story of the building of a cathedral tower, a tower which threatens to cause the collapse of the structure on which it rests. Priest and master builder confront each other, and the construction is accompanied by their dialogue, the dialogue between faith and technology. At one point the priest addresses the master builder in these words:

> My son. The building is a diagram of prayer; and our spire will be a diagram of the highest prayer of all. God revealed it to me in a vision, his unprofitable servant. He chose me. He chooses you, to fill the diagram with glass and iron and stone, since the children of men require a thing to look at. D'you think you can escape? You're not in my net ... It's His. We can neither of us avoid this work. And there's another thing. I've begun to see how we can't understand it either, since each new foot reveals a new effect, a new purpose.

Not only cathedrals, but every great engineering work is an expression of motivation and of purpose which cannot be divorced from religious implications. This truth provides the engineer with what many would assert to be the ultimate existential experience.

I do not want to get carried away on this point. The age of cathedral building is long past. And, as I have already said, less than one-quarter of today's engineers are engaged in construction activities of any sort. But every manmade structure, no matter how mundane, has a little bit of cathedral in it, since man cannot help but transcend himself as soon as he begins to design and construct. As the priest of *The Spire* expresses it: "each new foot reveals a new effect, a new purpose."

In spite of the many ugly and tasteless structures that mar our cities and landscapes, public enthusiasm for building has survived relatively unscathed through the recent years of disenchantment with technology. The engineer, in company with architects, artists, and city planners, has kept alive the public faith in the potentiality for beauty, majesty, and spirituality in construction.

At a time when we are embarrassed to recall the grandiose

pronouncements of so many of our predecessors, the First Procla-
mation of the Weimar Bauhaus, dating from 1919, retains its
dignity and ability to inspire. It was the concept of architect Walter
Gropius that great art in building grew out of craftsmanship, was
in fact nothing other than craftsmanship inspired. His concept of
craftsmanship included necessarily the essentials of civil engineering. "We must all turn to the crafts," he told his followers:

> Art is not a "profession." There is no essential difference
> between the artist and the craftsman. The artist is an exalted
> craftsman. In rare moments of inspiration, moments beyond
> the control of his will, the grace of heaven may cause his work
> to blossom into art. *But proficiency in his craft is essential to
> every artist.* Therein lies a source of creative imagination.
> Let us create a *new guild of craftsmen*, without the class
> distinctions which raise an arrogant barrier between craftsman
> and artist. Together let us conceive and create the new build-
> ing of the future, which ... will rise one day toward heaven
> from the hands of a million workers like the crystal symbol of a
> new faith.

Enough, then, of the civil engineer and his wrestling with the
elements, his love affair with nature, his yearning for immensity,
his raising toward heaven the crystal symbol of a new faith. His
existential bond to the earth, and expression of his own elemental
being, need no further amplification, no additional testimonials.

2.7 Decision-Making in the Corps of Engineers: The B. Everett Jordan Lake and Dam

P. Aarne Vesilind

A few years ago the U.S. Army Corps of Engineers officers started wearing a big button on their uniforms. The highly visible non-regulation insignia proclaimed: THE CORPS CARES.

There is no doubt that the officers of the Corps care. They are people of high moral standard and take their responsibility to the public seriously. Why then are the environmental groups continually fighting the actions of the Corps? Why then did the late Justice Douglass entitle a widely-read essay on the Corps "The Public Be Damned"? Why then is the Corps considered the diligent destroyer and not the protector of our environment?

This seeming incongruity is illustrated below in the article describing the decision-making with regard to the B. Everett Jordan Dam and Lake. As Samuel Florman suggests, it is indeed fun to build a dam. This enjoyment, coupled with the understandable desire to advance and be promoted within the Corps is much stronger than the concern for the environment. The Corps may indeed care, but for what?

THE HAW RIVER, a major tributary of the Cape Fear River, cascades down the fall line above Fayetteville, North Carolina and has been known to cause serious flooding in that eastern North

Carolina city. The flood of 1945 was particularly serious, causing over 2 million dollars worth of damage. Following a specific request by the people of Fayetteville, channeled through Senator Kerr Scott, the U.S. Army Corps of Engineers instituted a study of alternatives for flood protection.

The conventional solution to problems of flooding is to dam the offending river and thus capture the floodwaters behind the dam. In this case, however, the engineers encountered a problem. The Haw, at a point far enough upstream to assist Fayetteville, drops too rapidly and thus affords poor dam sites. The solution to this dilemma was ingenious: build a dam which captures most of the water in the Haw, but store it in a lake formed by a minor tributary, the New Hope Creek, forming a two-pronged lake (see the map on the next page). Following congressional approval, the construction of the New Hope Dam commenced in 1967.[1]

The first outspoken opponent of the project was Mr. Harold Cooley, Member of Congress, who staked his reelection campaign on this issue. Although he was stoutly supported by the farmers of Chatham County who stood to lose prime farm land to the lake, he lost in the City of Raleigh, and was defeated. At that time, the most vigorous supporter of the dam was Senator B. Everett Jordan, who owned a textile mill on the Haw River, twenty miles above where the lake would reach. Senator Jordan's non-support of Mr. Cooley contributed to the election of Mr. James Garner, a Republican who supported the project and served for one term in Congress.

The next public opposition to the project emerged from North Carolina State University, where E.H. Weiser, a hydrologist, had calculated the flood probabilities and had concluded that the Corps of Engineers' flood damage projections were grossly inflated.[2] Following the wide publication of this information, the North Carolina Conservation Council, a public interest organization, asked that the Corps prepare an Environmental Impact Statement (EIS) as required in the just-enacted National Environmental Policy Act. The Corps obliged by publishing an EIS in May, 1971, and a supplement was added in 1976. Although the EIS is theoretically supposed to be prepared as a planning document in order to judge

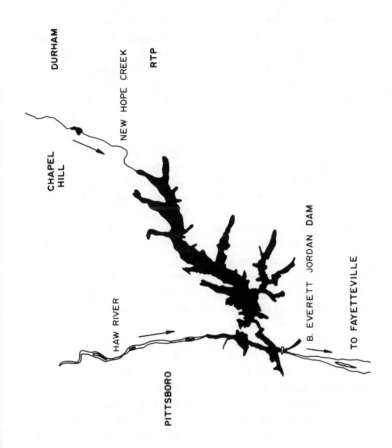

the feasibility and wisdom of initiating a specific project, the writing of the EIS and subsequent court battles did little to hinder the clearing of the land and the dam construction.

Following the publication of the court-directed supplement to the EIS, vigorous opposition developed to the project, based especially on the quality of water to be impounded. During this time, an independent benefit/cost study was conducted at the University of North Carolina by graduate students in environmental sciences and engineering. They found that a realistic calculation yielded a benefit/cost ratio of 0.3, where 1.0 would be required to justify the project.[3] It was also argued convincingly, by nearly all of the expert water quality engineers and scientists at the University of North Carolina, North Carolina State University and Duke University, that the lake would destroy thousands of acres of prime agricultural land, and that the water would be of questionable quality. Almost all of the experts agreed with this writer's conclusion that "If we looked for the absolute worst place to build a dam in North Carolina, we would not do much better than this site."[4]

Much of the water quality controversy centered on phosphorus. An analysis of the water which would flow into the lake showed that the level of nutrients was at least an order of magnitude higher than what would be necessary for accelerated eutrophication.[5] The high residence time in the New Hope arm of the lake, coupled with its high nutrient loading and shallow depth made the use of this water for recreation or other related uses highly questionable. Clearly phosphorus removal would be required by the towns of Durham and Chapel Hill, a cost not included in the Corps of Engineers benefit/cost analysis.

Responding to public pressure, the Corps decided to hire an independent disinterested consultant to establish once and for all the acceptability of the water quality. Hydrocomp, a respected hydrologic and water resources consuting firm from Palo Alto, California, was hired by the Corps to do the study.

The Hydrocomp report was published in 1976, as a supplement to the supplemental EIS, and showed conclusively that the water quality in much of the lake would be far below what the Corps had predicted, and that the New Hope arm of the lake would probably have serious water quality problems.[6]

A decision was necessary. Should the Corps continue insisting, as before, that the lake was needed (as it clearly was not, based on the benefit/cost analysis), and that the water quality would be acceptable (as it would clearly not be, based on the study funded by the Corps), or should the wisdom of completing the project and filling the lake be reevaluated? Which of these options would be chosen?

At this point in the chronology, I will digress to describe the history, function and operation of the Corps of Engineers in order to establish the framework in which this decision was made. I will then return to the question of the dam and show how the decision by the Corps is influenced by factors other than technical and economic considerations, or even the welfare of the people the Corps is supposed to serve.

the corps of engineers

The history of the U.S. Army Corps of Engineers stretches to the American Revolution, with the present Corps of Engineers tracing back to 1802 when it was formed by an act of Congress. Although originally meant to perform military duties, the Corps was instructed by Congress in 1824 to perform civilian duties as well, such as clearing snags from rivers. Since that time, the Corps' civilian duties have increased to projects totaling over 2.5 billion dollars per year.

The Corps has an impeccable tradition, with a reputation as a dedicated and honest civil servant. It consists of a cadre of about 200 regular Army officers who maintain a high morale among about 30,000 civilian employees. Typically, the career of a Corps officer begins at West Point, where only the brightest and best students receive appointments to the Corps.[7] Assignments at various districts and at headquarters in Washington follow, with retirement as a colonel or general. The closeness and intimacy of this elite group of officers contributes greatly to the efficiency and effectiveness of the Corps. It is, in short, a select club of highly skilled professionals, beholden to each other and to the Corps.[8]

The Corps of Engineers is, however, the only federal agency that doesn't follow the rules of executive/legislative conduct. Instead of incorporating its budget requests with the remainder of the administration plans, it reports through the Office of Management and Budget directly to Congress, thus bypassing scrutiny by other agencies and even the President. The Secretary of the Army, who is the titular head of the Corps, has only limited power to interfere in the Corp's activities since the financial power comes from Congress.

The Corps operates through the Public Works Committees of both the House and Senate. Under Public Law 82-298 the Chief of Engineers has special permission to spend up to $4,000,000 on any continuing authority project with only Congressional committee approval, and can initiate projects and provide a momentum which will be difficult to stem, since the pressure to finish a project on which millions have already been spent is overwhelming. (*Ed. Note:* At this writing, this authority has been temporarily suspended.) In effect, therefore, the Corps can dictate its own projects without even full Congressional oversight, much less any control from the executive branch. Not only does the Corps have power in Congress, but it farms out much of its budget as "research and development" to other federal agencies. This is welcome money, since the agencies do not have to struggle through Congress to obtain it, and thus few federal agencies are openly critical of the Corps.'

Every President since Franklin Roosevelt has tried, unsuccessfully, to either dismantle or curtail the powers of the Corps of Engineers. The Hoover Commission Task Force on Water Resources and Power recommended in 1949 that the civil functions of the Corps be transferred to the proposed Department of Natural Resources. No action was taken. In 1966, a bill sponsored by three senators to transfer some of the functions of the Corps died in committee. Even Lyndon Johnson, probably the best friend the Corps ever had, tried to keep it in line but failed. The last serious attempt to stifle the power of the Corps was by President Jimmy Carter when he recommended that over 40 Corps projects be abandoned as too costly and destructive to the environment. The political clout of the Corps was such, however, that only two were eventually stopped; the remainder continued as originally planned.

The Corps of Engineers has divided the United States into districts, usually based on river systems. A colonel is in command at each district, and the command is rotated every three years. The district commander is in direct charge of all of the projects in his district.

The Wilmington District office was, in 1976, headed by Colonel Homer Johnstone, who, following normal rotation, became the eighth engineer in charge of the New Hope Dam project, which by this time had been named in honor of its most ardent supporter, Senator B. Everett Jordan.

It was Colonel Johnstone who received the report from Hydrocomp which clearly showed that the dam had been a mistake. The construction of the dam was almost complete, however, and an admission that the project was indeed ill-considered might have significant repercussions. Colonel Johnstone had to make a decision: Should he proceed with the project and eventually fill the lake, or propose some other alternative?

possible options for the b. everett jordan dam

- Stop all further construction and abandon the project.
- Finish the project but leave the lake-bed dry. This would provide for maximum flood protection downriver, since the largest possible volume of water could be retained during floods, but benefits such as recreation and water supply would not be realized. This alternative would also save money, since some of the road construction necessary to elevate roads to above flood level (construction which extended well into the 1980s) would not be necessary.
- Finish the project as planned and fill the lake.

Although all disinterested expert opinion and formal testimony counseled against filling the lake, and choosing alternative 2, such a choice would in effect be an admission that the dam was a mistake. And if Colonel Johnstone admitted that the Corps had made a mistake, he would be indirectly criticizing his 7 predecessors (some of whom may have by then been generals in the Pentagon). What would that do to his career, and to the image of the Corps? Thus the decision became more than a technical or economical decision. Given the spirit of the Corps, the close-knit comeraderie and "old-boy" method of promotion, deciding that seven of one's predecessors (and presumably superiors) were wrong in their analysis of this project would have been professional suicide. Further, the image and reputation of the Corps as a monolithic and technically infallible organization would have been compromised. And finally, what would a decision to alter the project have done to the many politicians who had stoutly supported the original decision? It would have implied that they had bet on the wrong horse, and they would have concluded that the Corps had let them down — a notion which might have been reflected in the appropriations bill working its way through Congress.

what really happened?

On the day the Hydrocomp report was released to the public, the Corps accompanied it with a press release which had as its headline "WATER QUALITY IN JORDAN LAKE TO BE BETTER THAN ANTICIPATED." The story went on to say that although the report showed that much of the lake would probably not be useful for recreation and water supply, it would not be a "cesspool." The straw man tactic worked; the newspapers printed the headline verbatim, and the lake is filled. At the present time (1986) the towns of Chapel Hill and Durham are being asked to spend millions of dollars for phosphate removal processes at their wastewater treatment plants.

In all fairness to Colonel Johnstone, he may in fact have concluded, even in the face of overwhelming technical and public opinion, that the just and proper course of action was to continue with the project. We will never know what influenced the Colonel's decision. What we *do* know is that Colonel Johnstone went from Wilmington to Korea and became a brigadier general before being honorably retired from the Corps of Engineers.

references

1. "New Hope Lake," pamphlet prepared by the U.S. Army Corps of Engineers, Wilmington District, 1970.
2. Weiser, E., North Carolina State University, Raleigh, N.C., private communications.
3. "The New Hope Project-A Reevaluation," report by students at the Department of Environmental Sciences and Engineering, University of North Carolina in Chapel Hill, N.C., 1974.
4. Transcript of trial, North Carolina Conservation Council vs. Corps of Engineers, U.S. District Court, Greensboro, N.C.
5. Weiss, C. "Water Quality in the Haw River and the New Hope Creek," Water Resources Research Institute, North Carolina State University, Raleigh, N.C., 1973.
6. "Supplement to the Final Supplemental Environmental Impact Statement on the B. Everett Jordan Dam and Lake," Hydrocomp, Palo Alto, Cal., 1976.
7. Lenny, J.J. *Caste System in the American Army: A Study of the Corps of Engineers and Their West Point System*, Greenberg Publishing, New York, 1949.
8. Laycock, G. *The Diligent Destroyers*, Audobon/Ballentine, New York, 1970.
9. Douglas, Justice William O. "The Corps of Engineers: The Public Be Damned," *Playboy*, July, 1969.
10. *The Chapel Hill Newspaper*, Chapel Hill, N.C.

2.8 Moral Development and Professional Engineering

Elizabeth M. Endy
and P. Aarne Vesilind

Why is there such a divergence of opinion among engineers when they are faced with ethical dilemmas? Why do some engineers react in ways which others would find either repugnant, or stupid? And do these opinions change as the engineer progresses through his/her career? Would a young second lieutenant in the Corps of Engineers make the same decision as the colonel? Or is it possible that time, and the confrontation of dilemmas where values conflict, cause one to grow and move into more reasoned and less dogmatic moral positions?

While reading the article below, try to place real people with whom you have had professional contact into the various stages of moral development in engineering. Then consider how your own decisions could have been classified.

Perhaps a word of caution is appropriate. In the article below, qualitative material is managed in a quantitative way. As with any such enumeration, considerable argument can arise as to the accuracy of the criteria used. This is a valid criticism. Further, the qualitative material is listed in an hierarchical manner, implying that one end of the scale is better than the other. This implication is not intended. Value judgments as to what is the better stage of moral development are left to the reader.

This article, in revised form, first appeared in *Civil Engineering*, December 1985. Used with permission of the American Society of Civil Engineers.

"WHAT DO YOU mean I won't get paid?"

The consulting engineer is understandably outraged. He has spent a week evaluating the design concept for a solid waste processing facility. His review for EPA is complete; he has done his duty as a disinterested evaluator. His value engineering effort has discovered that the cost of the facility has been greatly underestimated. And now the design engineer is saying that his firm cannot pay for a negative review. Only if the project is approved and funded by EPA will he get paid. The consulting engineer must either forfeit a week's pay, or rewrite his review as an approval of the project design.

The engineer has become involved in an ethical dilemma. If he does what he considers "right" (to present the facts), he will cause material harm to himself: he won't get paid. In order to act within this dilemma, the engineer must apply some ethical system of problem-solving.

Each person has, during his/her life and career, developed these ethical standards, but where do these ethics come from, and what factors influence the choice of personal ethical standards?

In addition to their responsibilities as citizens, professional engineers, if they are to fulfill completely their obligations to society, must also recognize the ethical dimensions of their professional duties. The moral values which provide the foundation for ethical conduct develop with experience in each individual, starting at an early age with the recognition of the value of "good." How these values are translated into professional practice determines how each individual acts in a professional situation like the one described above, where ethics may influence the selection of alternative courses of action.

Research in psychology has shown that the development of moral cognition can be quantified and scored. The objective of this paper is to illustrate the parallels between moral development and the development of professional engineering ethics.

moral development

At about two years of age, children begin to acquire a sense of what is right and what is wrong. They begin to understand that some of their friends are "nice" and "good" while others are not. This elementary understanding of fairness and justice continues to develop into increasingly more sophisticated outlooks on problems of social interaction. *Moral development* is the growth from early childhood, into and throughout adulthood, of a person's abilities

1. to identify those situations in which decisions and actions are indeed based on some concepts of justice, rightness, duty, and caring,
2. to reason toward a choice of action in such a situation, and
3. to follow through on that choice with appropriate action.

In 1932, Jean Piaget found that children progress through well-defined stages in their development of the concept of fairness, and that these stages are measurable.[1] It should be possible, according to Piaget, to ask the proper questions and thus to determine the stage of this development in a child.

Whereas Piaget studied young children, aged 6 to 11, Lawrence Kohlberg extended this research to adolescents and young adults, using interviews every 3 or 4 years over a 20-year longitudinal study. Kohlberg presented each subject with hypothetical moral dilemmas and then, using a structured scoring system, he identified six stages of moral development as revealed in the subjects' reasoning.[2] The criteria for these developmental stages are:

1. Each stage reveals qualitatively distinct modes of thinking, called "cognitive structures." These cognitive structures are the underlying organization of thought, and are not acquired skills or knowledge.
2. Individuals develop from lower to higher stages in an invariant sequence; environmental factors may influence the speed of the development but not the direction of it.
3. Stages are not skipped over. For example, a person reasoning at Stage 3 is capable of reasoning at Stages 1 and 2 and, if he/she is in transition, at Stage 4, but not at Stage 5.

4. Development from one stage to another occurs when the individual encounters an experience that does not fit into his/her present cognitive structures. Through assimilating the experience, the person develops new structures of thought, thus leading him/her into the next stage.

Kohlberg's data led him to organize six Piagetian stages of moral development, in three levels. These are outlined below, based on summaries by Rest[3] and Colby.[4]

LEVEL 1   Preconventional
Stage 1: Obedience — we do what we are told in order to avoid punishment. At this stage, the child is obedient to the caretaker, and being good is simply to do as told. The child sees no plan or purpose to the rules; they are simply to be obeyed.

Stage 2: Purposeful exchange — we'll make a deal. There is a realization that our doing what others want may result in *their* doing what we want. If, for example, we don't scream, we may get a lollipop.

LEVEL 2   Conventional
Stage 3: Being a nice person — having motives acceptable by peers will be of personal benefit. Being cooperative, not only on a deal-to-deal basis, but as a style of relating to people, is recognized as beneficial to oneself, as well as to others. Stage 3 behavior often results in loyalty and trust in friendships.

Stage 4: Law and order — everyone should obey the law. If one simply follows the letter of the law, one is correct and "moral." Because the laws apply to all, it is easy to anticipate the actions of fellow citizens. This is a stable social system, because everyone is expected to conform. Many adults act on this level in terms of their moral judgments.

LEVEL 3   Postconventional or principled
Stage 5: Societal consensus — obey the laws of the majority. In this stage, the difference between a bad and good law is recognized, and good laws are defined as those which are developed by

the democratic process. As long as people's lives, liberties, and pursuit of happiness are guaranteed, and the laws are democratically developed as the laws of the majority, then to act morally is to follow these laws.

Stage 6: Recognition and acceptance of universal moral principles, such as equality of rights and dignity of all individuals. Most laws in democratic societies are based on such principles. If, however, a law does violate a universal principle, a person at Stage 6 will act according to the principle and not the law.

professional moral development

Professional engineers are required to interact with society not only as professionals who fulfill the wishes of society which has educated them to perform some specialized function, but also and perhaps foremost as human beings. Often professional decisions become muddled with personal morals. Thus it is reasonable to assume that the levels of moral development described above apply to engineers in their professional conduct as well.

What separates the professional engineer (as well as other professionals) from the rest of society is that by virtue of their training they have assumed a responsibility to society, and to the profession of which they are members. Thus a professional engineer has, in addition to personal ethics, a second layer of ethics which apply to his/her professional conduct.

Using this notion of the similitude between private moral development and professional moral development, Richard Mc-Cuen[5] has suggested six stages of professional development. The following is based on McCuen's ideas, but does not strictly parallel his interpretations of the stages of professional conduct:

LEVEL 1   Preprofessional

Stage 1: At this level, the engineer is not concerned with social or professional responsibilities. Professional conduct is dictated by the gain to the individual, with no thought of how such conduct would affect the firm, the client/engineer relationship, or the profession.

Stage 2: Just as the child recognizes, in Stage 2, that there is something to be gained by "being nice," the engineer functioning in this stage sees his/her conduct as affecting his/her marketability as a professional, and thus while the engineer is aware of the ideas of loyalty to the firm, client confidence and proper professional conduct, the ethical behavior is based on the motive of self-advancement.

LEVEL 2 Professional

Stage 3: At this stage, the engineer puts loyalty to the firm above any other consideration. Since he/she works for the firm, the firm can dictate the proper action, and the engineer is freed from further ethical considerations. Samuel Florman has recognized this freedom in engineering and publicized it as *The Existential Pleasures of Engineering.*[6] Just as the child recognizes that "nice" behavior is rewarded, so the engineer buries himself or herself in technical matters, becomes the so-called "team player" for the firm, and is "nice" by ignoring the ramifications of the job on society and on the environment.

Stage 4: At this stage, the individual retains loyalty to the firm, but recognizes that the firm is part of the larger profession, and thus loyalty to the profession enhances the reputation of the firm and brings rewards to the engineer. Engineering practice is viewed from a purely professional perspective, with no thought toward the larger issues of professional responsibility and social welfare. It is noted previously that many adults attain Stage 4 in their everyday moral conduct, where the laws of the land, regardless of how the laws were enacted, deserve their support. Engineers who follow the laws of professional conduct as dictated by the various professional societies are acting at this stage.

LEVEL 3 Principled Professional

Stage 5: Service to human welfare is considered paramount at Stage 5, and it is recognized that such service will also bring credit to the firm and to the profession. The driving force now is service to society, and thus it is the rules of society which must be recognized in determining professional conduct. Where professional standards do not apply or are in conflict with the prevailing morals of society, the latter takes precedence. As is the case with Kohlberg's stages, at this stage the validity of the social rules is not questioned, and as long as the rules are arrived at by democratic social consensus, they are accepted as the proper criteria for professional conduct.

Stage 6: At this stage, professional conduct is dictated by universal rules of justice, fairness and caring for fellow humans and the whole of nature. It is most difficult working at this level because it is possible to perform actions which, although correct in the universal sense of justice and caring, contradict the prevailing social order and/or the code of ethical conduct for the profession. For example, until 1977 the Code of Ethics for the American Society of Civil Engineers had a footnote indicating that, in effect, although bribes were not condoned in most countries, where bribes were a standard method of doing business, bribery of public officials was accepted engineering practice. An engineer working at any stage other than Stage 6 would have thus used bribes as standard operating procedure whenever working in a country where bribery was practiced as a matter of course.

Awareness of these stages in the development of professional ethics helps us to understand our behavior and the behavior of co-workers when we find ourselves faced with ethical decisions.

case studies involving ethics and professional conduct

The following case studies illustrate the usefulness of applying the model of staged moral development in clarifying the ethical aspects of engineering decisions.

CASE 1   Conflict of Interest
Large consulting firms commonly have many offices, and often communication among the offices is less than efficient. Engineer Stan, in the Atlanta office, is retained by a neighborhood association to write an environmental impact study which concludes that the plans by a private oil company to build a petrochemical complex would harm the habitat

of several endangered species. The client has already reviewed draft copies of the report and is planning to hold a press conference when the final report is delivered. A few days before the final report is presented to the client, the New York office becomes aware of the study and tells Stan to postpone its delivery. Engineer Bruce from New York flies down to explain to Stan that the oil company is one of the firm's most valued clients, and that the president of the oil company has found out about Stan's report and threatens to pull all of their business should the report be delivered to the neighborhood association. Bruce tells Stan to rewrite the report in such a way as to show that there would be no significant damage to the environment.

At what stage is Engineer Bruce acting? Obviously, he does not have social welfare in the forefront of his professional conduct, and is putting the firm and its business ahead of any social concern. He would be acting at Stage 2 or 3. If Engineer Stan does as he is told, he is simply following orders and thus working at Stage 2 or 3. But Stan has other options. He could decide that he cannot rewrite the report since this would put him in very bad light with his clients, who have already seen the draft copy. He can therefore disengage himself from the job and return the client's payments, invoking (correctly) a conflict of interest. He would then put his loyalty to the profession in the forefront, and would be working at Stage 4. Finally, Stan could refuse to rescind his report, and be fired. Stage 5 behavior would allow him to simply drop the whole matter, whereas if he felt strongly that the neighborhood group was speaking for the environment and that they badly needed his services, he could continue working with them, and risk the animosity of his former colleagues in the firm. The latter would be Stage 6 behavior, — putting universal principles above all other considerations.

CASE 2   Loyalty to the Employer
Engineer Diane works for a large international consulting firm which has been retained by a third-world nation to assist in the construction of a water supply reservoir. She soon discovers that the conditions for the workers are appalling, and that several serious accidents occur every

day, with a number of deaths. She tries to get the contractor to correct the problem, but is told to mind her own business. She receives the same answer from the government which is paying her salary. One day a delegation of the workers visits her in secret, tells her that they plan to go on strike the next day and begs her to join them in the picket lines for moral support.

The ASCE Code of Ethics clearly discourages engineers from taking part in any strikes or other type of work stoppages. If Engineer Diane is acting at Stage 4 or below, she would not take part in the walkout. At Stage 5, she would note that she works in a society which condones such treatment of workers, and even though she might see the injustice in the system, she still would not join the workers because they happened to live in that country. Had the same treatment occurred in the USA, she probably would have agreed to join in the walkout. And finally, if her professional development is at Stage 6, she would recognize the overriding justice in the workers' demands and join them. This would probably cost her her job.

CASE 3   Unsafe Design

Engineer Jody, fresh out of school, is asked to design reinforcing bars for a cantilever beam which is to be part of an entranceway for a hospital. Her superior, Engineer Robert, looks over her calculations, without actually checking them, and passes the drawings over to the drafting department. The plans are drawn and sealed by Engineer Robert, and the beam poured. Just by chance, Engineer Pat discovers a serious error in the plans, and points it out to Engineer Robert. There apparently are one half as many reinforcing bars in the beam as good design would dictate. Engineer Robert notes that the beam seems to be holding, and instructs the drafting department to change the drawings by adding the missing reinforcing bars to the "as built" drawings.

Clearly Engineer Robert is acting unethically, perhaps at Stage 1, since his action would not help him even personally in the future. But what should Engineer Pat do? The simplest option would be to forget the whole thing (Stage 3 or below). She could go to a superior in the firm, or even report the incident to the professional licensing board. This would be action at Stage 4. The action of the registration board would then dictate what Engineer Pat would do next. If she felt strongly that the structure was unsafe, she could go public with her concerns. This action would be very much against the codes of professional conduct, which specifically forbid criticizing the work of another engineer in public. But if this is Engineer Pat's only recourse — if the structural defect is not to be corrected, and if this would, in her opinion, present a clear danger to the public, Stage 6 reasoning would dictate a carefully considered but forceful statement to the public media.

conclusion

One criticism of Piaget and Kohlberg is that they measure moral development on the basis of verbal responses to structured moral dilemmas. It has been shown by numerous researchers, however, that what one *thinks* about a situation is often quite different from how one *acts* in a similar predicament. In judging one's own personal conduct, it is necessary to consider what one *does* in professional practice as well as what one *might* do in a given circumstance. In engineering, as in every aspect of life, the ultimate test of moral development is manifested in *actions*.

references

1. Piaget, J. *The Moral Judgement of the Child* (M. Gabian, trans.), The Free Press, New York, 1965 (originally published in 1932).
2. Kohlberg, L. *Collected Papers on Moral Development and Moral Education*, Laboratory of Human Development, Harvard University, Cambridge MA, 1973.
3. Rest, J.R. *Developments in Judging Moral Issues*, University of Minnesota Press, Minneapolis, 1979.
4. Colby, A., L. Kohlberg, J. Gibbs, and M. Lieberman "Longitudinal Study of Moral Development," *Monograph of the Society for Research in Child Development*, v. 48 (1, serial 200) Univ. of

Chicago Press, Chicago IL, 1983.
5. McCuen, R.H. "The Ethical Dimensions of Professionalism," *Journal of Professional Activities, ASCE* v. 105, n. E12, April 1979.
6. Florman, S. *The Existential Pleasures of Engineering*, St. Martin's Press, New York, 1976.

2.9 Should Trees Have Standing?

Christopher D. Stone

In the introduction to this essay, Garrett Hardin writes: "Law, to be stable, must be based on ethics. In evoking a new ethic to protect land and other natural amenities, [Aldo] Leopold implicitly called for concomitant changes in the philosophy of the law. Now, less than a generation after the publication of Leopold's classic essay, Professor Christopher D. Stone has laid the foundation for just such a philosophy in a graceful essay that itself bids fair to become a classic."

Reproduced below is the "Introduction" to the essay, in which the basic framework for the argument is established. The essay was first written for Southern California Law Review.

IN *THE Descent of Man,* written a full century ago, Charles Darwin observed that the history of man's moral development has been a continual extension in the range of objects receiving his "social instincts and sympathies." Originally each man had moral concern only for himself and those of a very narrow circle about him; later, he came to regard more and more "not only the welfare, but the happiness of all his fellow men." Then, gradually, "his sympathies became more tender and widely diffused, extending to men of all races, to the imbecile, maimed and other useless members of society, and finally to the lower animals...."

The history of the law suggests a parallel development. The scope of "things" accorded legal protection has been continuously extending. Members of the earliest "families" (including extended kinship groups and clans) treated everyone on the outside as suspect, alien, and rightless, except in the vacant sense of each man's "right to self-defense." "An Indian Thug," it has been written, "conscientiously regretted that he had not robbed and strangled as many travelers as did his father before him. In a crude state of civilization the robbery of strangers is, indeed, generally considered as honorable." And even within a single family, persons we presently regard as the natural holders of at least some legal rights had none. Take, for example, children. We know something of the early rights-status of children from the widespread practice of infanticide — especially of the deformed and female. (Senicide, practiced by the North American Indians, was the corresponding rightlessness of the aged.) Sir Henry Maine tells us that as late as the *Patria Potestas* of the Romans, the father had *jus vitae necisque* — the power of life and death — over his children. It followed legally, Maine writes, that

> he had power of uncontrolled corporal chastisement; he can modify their personal condition at pleasure; he can give a wife to his son; he can give his daughter in marriage; he can divorce his children of either sex; he can transfer them to another family by adoption; and he can sell them.

The child was less than a person: it was, in the eyes of the law, an object, a thing.

The legal rights of children have long since been recognized in principle, and are still expanding in practice. Witness, just within recent time, *In re Gault*, the United States Supreme Court decision guaranteeing basic constitutional protections to juvenile defendants, and the Voting Rights Act of 1970, with its lowering of the voting age to eighteen. We have been making persons of children although they were not, in law, always so. And we have done the same, albeit imperfectly some would say, with prisoners, aliens, women (married women, especially, were nonpersons through most of legal history), the insane, blacks, fetuses, and Indians.

People are apt to suppose that there are natural limits on how far the law can go, that it is only matter in human form that can come to be recognized as the possessor of rights. But it simply is not so. The world of the lawyer is peopled with inanimate right-holders: trusts, corporations, joint ventures, municipalities, Sub-chapter R partnerships, and nation-states, to mention just a few. Ships, still referred to by courts in the feminine gender, have long had an independent jural life, often with striking consequences. In one famous U.S. Supreme Court case a ship had been seized and used by pirates. After the ship's capture, the owners asked for her return; after all, the vessel had been pressed into piracy without their knowledge or consent. But the United States condemned and sold the "offending vessel." In denying release to the owners, Justice Story quoted Chief Justice Marshall from an earlier case:

> This is not a proceeding against the owner; it is a proceeding against the vessel for an offense committed by the vessel; which is not the less an offense...because it was committed without the authority and against the will of the owner.

The ship was, in the eyes of the law, the guilty person.

We have become so accustomed to the idea of a corporation having "its" own rights, and being a "person" and "citizen" for so many statutory and constitutional purposes, that we forget how perplexing the notion was to early jurists. "That invisible, intangible and artificial being, that mere legal entity," Chief Justice Marshall wrote of the corporation in *Bank of the United States v. Deveaux* — could a suit be brought in its name? Ten years later, in the *Dartmouth College* case, he was still refusing to let pass unnoticed the wonder of an entity "existing only in contemplation of law." Yet, long before Marshall worried over the personification of the modern corporation, the best medieval legal scholars had spent hundreds of years struggling with the legal nature of those great public "corporate bodies," the Church and the State. How could they exist in law, as entities transcending the living pope and king?

It was clear how a king could bind himself — on this honor — by a treaty. But when the king died, what was it that was burderned with the obligations of, and claimed the rights under, the treaty his tangible hand had signed? The medieval mind saw (what we have lost our capacity to see) how unthinkable it was, and worked out the most elaborate conceits and fallacies to serve as anthropomorphic flesh for the Universal Church and the Universal Empire.

It is this note of the unthinkable that I want to dwell upon for a moment. Throughout legal history, each successive extension of rights to some new entity has been, theretofore, a bit unthinkable. Every era is inclined to suppose the rightlessness of its rightless "things" to be a decree of Nature, not a legal convention — an open social choice — acting in support of some status quo. It is thus that we avoid coming face to face with all the moral, social, and economic dimensions of what we are doing. Consider, for example, how the United States Supreme Court sidestepped the moral issues behind slavery in its 1856 *Dred Scott* decision; blacks had been denied to rights of citizenship "as a suboridnate and inferior class of beings." Their unfortunate legal status reflected, in other words, not our choice at all, but "just the way things were." In an 1856 contest over a will, the deceased's provision that his slaves should decide between emancipation and public sale was held void on the ground that slaves had no legal capacity to choose. "These decisions," the Virginia court explained,

> are legal conclusions flowing naturally and necessarily from the one clear, simple, fundamental idea of chattel slavery. That fundamental idea is, that, in the eye of the law, so far certainly as civil rights and relations are concerned, the slave is not a person, but a thing. The investiture of a chattel with civil rights or legal capacity is indeed a legal solecism and absurdity. The attribution of a legal conscience, legal intellect, legal freedom, or liberty and power of free choice and action, and corresponding legal obligations growing out of such qualities, faculties and action — implies a palpable contradiction in terms.

In a like vein, the highest court in California once explained
that Chinese had not the right to testify against white men in
criminal matters because they were ''a race of people whom nature
has marked as inferior, and who are incapable of progress or
intellectual development beyond a certain point ... between whom
and ourselves nature has placed an impassable difference.''

The popular conception of the Jew in the thirteenth century
contributed to a law which treated them, as one legal commentator
has observed, as ''men *ferae naturae,* protected by a quasi-forest
law. Like the roe and the deer, they form an order apart.'' Recall,
too, that it was not so long ago that the fetus was ''like the roe and
the deer.'' In an early suit attempting to establish a wrongful death
action on behalf of a negligently killed fetus (now widely accepted
practice in American courts), Holmes, then on the Massachusetts
Supreme Court, seems to have thought it simply inconceivable
''that a man might owe a civil duty and incur a conditional
prospective liability in tort to one not yet in being.'' The first
woman in Wisconsin who thought she might have a right to
practice law was told that she did not. We had nothing against
them, of course; but they were *naturally* different.

> The law of nature destines and qualifies the female sex for the
> bearing and nurture of the children of our race and for the
> custody of the homes of the world ... lifelong callings of
> women, inconsistent with these radical and sacred duties of
> their sex, as is the profession of the law, are departures from
> the order of nature; and when voluntary, treason against it....
> The peculiar qualities of womanhood, its gentle graces, its
> quick sensibility, its tender susceptibility, its purity, its deli-
> cacy, its emotional impulses, its subordination of hard reasons
> to sympathetic feeling, are surely not qualifications for foren-
> sic strife. Nature has tempered woman as little for the juridical
> conflicts of the court room, as for the physical conflicts of the
> battlefield ...

The fact is, that each time there is a movement to confer rights
onto some new ''entity'' the proposal is bound to sound odd or

frightening or laughable.* This is partly because until the rightless thing receives its rights, we cannot see it as anything but a thing for the use of "us" — those who are holding rights at the time.

Thus it was that the Founding Fathers would speak of the inalienable rights of all men, and yet maintain a society that was, by modern standards, without the most basic rights for blacks, Indians, children and women. There was no hypocrisy; emotionally, no one felt that these other things were *men*. In this vein, what is striking about the Wisconsin case above is that the court, for all its talk about women, so clearly was never able to see women as they are and might become. All it could see was the popular "idealized" version of an object it needed. Such is the way the slave South looked upon the black. "The older South," W.E. Dubois wrote, clung to "the sincere and passionate belief that somewhere between men and cattle, God created a *tertium quid*, and called it a Negro."

Obviously, there is something of a seamless web involved: there will be resistance to giving a "thing" rights until it can be seen and valued for itself; yet, it is hard to see it and value a "thing" for itself until we can bring ourselves to give it rights — which is almost inevitably going to sound inconceivable to a large

---

*Recently, a group of prison inmates in Suffolk County tamed a mouse that they discovered, giving him the name Morris. Discovering Morris, a jailer flushed him down the toilet. The prisoners brought a proceeding against the warden complaining, *inter alia*, that Morris was subjected to discriminatory discharge and was otherwise unequally treated. The action was unsuccessful, the court noting that the inmates themselves were "guilty of imprisoning Morris without a charge, without a trial, and without bail," and that other mice at the prison were not treated more favorably. "As to the true victim, the Court can only offer again the sympathy first proffered to his ancestors by Robert Burns's poem, 'To a Mouse'."

The whole matter seems humorous, of course. But we need to know more of the function of humor in the unfolding of a culture, and the ways in which it is involved with the social growing pains to which it is testimony. Why do people make jokes about the Women's Liberation Movement? Is it not on account of — rather than in spite of — the underlying validity of the protests and the uneasy awareness that a recognition of the claims is inevitable? Arthur Koestler rightly begins his study of the human mind, *Act of Creation* (1964), with an analysis of humor, entitled "The Logic of Laughter." Cf. Freud's paper, "Jokes and the Unconscious."

group of people.

The reader must know by now, if only from the title of the book, the reason for this little discourse on the unthinkable. I am quite seriously proposing that we recognize legal rights of forests, oceans, rivers and other so-called "natural objects" in the environment — indeed, of the natural environment as a whole.

As strange as such a notion may sound, it is neither fanciful nor without considerable operational significance. In fact, I do not think it would be a misdescription of recent developments in the law to say that we are already on the verge of such an assignment of rights to nature, although we have not faced up to what we are doing in those particular terms.

We should do so now, and begin to explore the implications such an idea would yield.

Appendix

A.1   Code of Ethics   *American Society of Civil Engineers*

A.2   Primer on Ethical Theories

CODE OF ETHICS*

Effective January 1, 1977

FUNDAMENTAL PRINCIPLES**

Engineers uphold and advance the integrity, honor and dignity of the engineering profession by:

1. using their knowledge and skill for the enhancement of human welfare;

2. being honest and impartial and serving with fidelity the public, their employers and clients;

3. striving to increase the competence and prestige of the engineering profession; and

4. supporting the professional and technical societies of their disciplines.

FUNDAMENTAL CANONS

1. Engineers shall hold paramount the safety, health and welfare of the public in the performance of their professional duties.

2. Engineers shall perform services only in areas of their competence.

3. Engineers shall issue public statements only in an objective and truthful manner.

4. Engineers shall act in professional matters for each employer or client as faithful agents or trustees, and shall avoid conflicts of interest.

5. Engineers shall build their professional reputation on the merit of their services and shall not compete unfairly with others.

6. Engineers shall act in such a manner as to uphold and enhance the honor, integrity, and dignity of the engineering profession.

7. Engineers shall continue their professional development throughout their careers, and shall provide opportunities for the professional development of those engineers under their supervision.

---

*As adopted September 25, 1976 and amended October 25, 1980.
**The American Society of Civil Engineers adopted THE FUNDAMENTAL PRINCIPLES of the ABET Code of Ethics of Engineers as accepted by the Accreditation Board for Engineering and Technology, Inc. (ABET).
(By ASCE Board of Direction action April 12–14, 1975)

ASCE GUIDELINES TO PRACTICE UNDER THE FUNDAMENTAL CANONS OF ETHICS

CANON 1. Engineers shall hold paramount the safety, health and welfare of the public in the performance of their professional duties.

a. Engineers shall recognize that the lives, safety, health and welfare of the general public are dependent upon engineering judgments, decisions and practices incorporated into structures, machines, products, processes and devices.

b. Engineers shall approve or seal only those design documents, reviewed or prepared by them, which are determined to be safe for public health and welfare in conformity with accepted engineering standards.

c. Engineers whose professional judgment is overruled under circumstances where the safety, health and welfare of the public are endangered, shall inform their clients or employers of the possible consequences.

d. Engineers who have knowledge or reason to believe that another person or firm may be in violation of any of the provisions of Canon 1 shall present such information to the proper authority in writing and shall cooperate with the proper authority, in furnishing such further information or assistance as may be required.

e. Engineers should seek opportunities to be of constructive service in civic affairs and work for the advancement of the safety, health and well-being of their communities.

f. Engineers should be committed to improving the environment to enhance the quality of life.

CANON 2. Engineers shall perform services only in areas of their competence.

a. Engineers shall undertake to perform engineering assignments only when qualified by education or experience in the technical field of engineering involved.

b. Engineers may accept an assignment requiring education or experience outside of their own fields of competence, provided their services are restricted to those phases of the project in which they are qualified. All other phases of such project shall be performed by qualified associates, consultants, or employees.

c. Engineers shall not affix their signatures or seals to any engineering plan or document dealing with subject matter in which they lack competence by virtue of education or experience or to any such plan or document not reviewed or prepared under their supervisory control.

CANON 3. Engineers shall issue public statements only in an objective and truthful manner.

a. Engineers should endeavor to extend the public knowledge of engineering, and shall not participate in the dissemination of untrue, unfair or exaggerated statements regarding engineering.

b. Engineers shall be objective and truthful in professional reports, statements, or testimony. They shall include all relevant and pertinent information in such reports, statements, or testimony.

c. Engineers, when serving as expert witnesses, shall express an engineering opinion only when it is founded upon adequate knowledge of the facts, upon a background of technical competence, and upon honest conviction.

d. Engineers shall issue no statements, criticisms, or arguments on engineering matters which are inspired or paid for by interested parties, unless they indicate on whose behalf the statements are made.

e. Engineers shall be dignified and modest in explaining their work and merit, and will avoid any act tending to promote their own interests at the expense of the integrity, honor and dignity of the profession.

CANON 4. Engineers shall act in professional matters for each employer or client as faithful agents or trustees, and shall avoid conflicts of interest.

a. Engineers shall avoid all known or potential conflicts of interest with their employers or clients and shall promptly inform their employers or clients of any business association, interests, or circumstances which could influence their judgment or the quality of their services.

b. Engineers shall not accept compensation from more than one party for services on the same project, or for services pertaining to the same project, unless the circumstances are fully disclosed to and agreed to, by all interested parties.

c. Engineers shall not solicit or accept gratuities, directly or indirectly, from contractors, their agents, or other parties dealing with their clients or employers in connection with work for which they are responsible.

d. Engineers in public service as members, advisors, or employees of a governmental body or department shall not participate in considerations or actions with respect to services solicited or provided by them or their organization in private or public engineering practice.

e. Engineers shall advise their employers or clients when, as a result of their studies, they believe a project will not be successful.

f. Engineers shall not use confidential information coming to them in the course of their assignments as a means of making personal profit if such action is adverse to the interests of their clients, employers or the public.

g. Engineers shall not accept professional employment outside of their regular work or interest without the knowledge of their employers.

CANON 5. Engineers shall build their professional reputation on the merit of their services and shall not compete unfairly with others.

a. Engineers shall not give, solicit or receive either directly or indirectly, any commission, political contribution, or a gift or other consideration in order to secure work, exclusive of securing salaried positions through employment agencies.

b. Engineers should negotiate contracts for professional services fairly and on the basis of demonstrated competence and qualifications for the type of professional service required.

c. Engineers shall not request, propose or accept professional commissions on a contingent basis under circumstances in which their professional judgments may be compromised.

d. Engineers shall not falsify or permit misrepresentation of their academic or professional qualifications or experience.

e. Engineers shall give proper credit for engineering work to those to whom credit is due, and shall recognize the proprietary interests of others. Whenever possible, they shall name the person or persons who may be responsible for designs, inventions, writings or other accomplishments.

f. Engineers may advertise professional services in a way that does not contain self-laudatory or misleading language or is in any other manner derogatory to the dignity of the profession. Examples of permissible advertising are as follows:

Professional cards in recognized, dignified publications, and listings in rosters or directories published by responsible organizations, provided that the cards or listings are consistent in size and content and are in a section of the publication regularly devoted to such professional cards.

Brochures which factually describe experience, facilities, personnel and capacity to render service, providing they are not misleading with respect to the engineer's participation in projects described.

Display advertising in recognized dignified business and professional publications, providing it is factual, contains no laudatory expressions or implication and is not misleading with respect to the engineer's extent of participation in projects described.

A statement of the engineers' names or the name of the firm and statement of the type of service posted on projects for which they render services.

Preparation or authorization of descriptive articles for the lay or technical press, which are factual, dignified and free from laudatory implications. Such articles shall not imply anything more than direct participation in the project described.

Permission by engineers for their names to be used in commercial advertisements, such as may be published by contractors, material suppliers, etc., only by means of a modest, dignified notation acknowledging the engineers' participation in the project described. Such permission shall not include public endorsement of proprietary products.

g. Engineers shall not maliciously or falsely, directly or indirectly, injure the professional reputation, prospects, practice or employment of another engineer or indiscriminately criticize another's work.

h. Engineers shall not use equipment, supplies, laboratory or office facilities of their employers to carry on outside private practice without the consent of their employers.

CANON 6. Engineers shall act in such a manner as to uphold and enhance the honor, integrity, and dignity of the engineering profession.

a. Engineers shall not knowingly act in a manner which will be derogatory to the honor, integrity, or dignity of the engineering profession or knowingly engage in business or professional practices of a fraudulent, dishonest or unethical nature.

CANON 7. Engineers shall continue their professional development throughout their careers, and shall provide opportunities for the professional development of those engineers under their supervision.

a. Engineers should keep current in their specialty fields by engaging in professional practice, participating in continuing education courses, reading in the technical literature, and attending professional meetings and seminars.

b. Engineers should encourage their engineering employees to become registered at the earliest possible date.

c. Engineers should encourage engineering employees to attend and present papers at professional and technical society meetings.

d. Engineers shall uphold the principle of mutually satisfying relationships between employers and employees with respect to terms of employment including professional grade descriptions, salary ranges, and fringe benefits.

A.2 Primer on Ethical Theories

AN ETHICAL THEORY is an attempt to answer certain questions about ethics. Ethical theories do not directly provide answers to specific questions such as whether prayer should be allowed in public schools or how much to donate to charity. Rather, they are attempts to decide what moral principles are correct. At a more abstract level, known as "meta-ethics," philosophers attempt to analyze key ethical concepts such as "rights" or "duty." Ethical theories are used as input in the development of a code or other decision making procedures — "applied ethics."

two types of ethical theory

In the Western tradition, ethical theories are of two main types. *Consequentialist* (or *teleological*) theories evaluate acts, policies, practices and institutions according to their consequences. Roughly speaking, in such theories a right action is one which has good consequences; a wrong action is one which has bad consequences. The most influential consequentialist theory is *utilitarianism*, developed in great part by Jeremy Bentham (1748-1832), who held the view that only happiness is good in itself; an act is right in proportion as it tends to increase the sum of human happiness or decrease unhappiness. Some philosophers have tried to extend utilitarianism to animals, as seen in Chapter 1.2. Most current decision strategies (game theory, decision theory, cost/benefit analysis, risk/benefit analysis, and other derivatives of operations research) trace their ethical origin to utilitarianism. Other consequentialist ethical theories locate the "good" elsewhere: for example, *ethical egoism* (the good is that which benefits me), or *nationalism* (the good is that which advances the state).

Deontological theories are non-consequentialist; deontologists deny that the rightness or wrongness of acts or rules is reducible to the value of their consequences. Any ethical theory which presents morality as a system of absolute duties to do or forbear from certain actions, is a deontological system; for example, the Ten Commandments. Deontologists do not deny that it is desirable to bring about good ends, of course, but they do regard a number of things as good or evil regardless of consequences. The German philosopher Immanuel Kant (1724-1804) emphasizes the absolute value of persons, who as free, rational beings must always be treated as ends in themselves. To act morally is to follow universal moral principles which require respect for persons. Few deontological theories, however, regard ethical behavior as the mere slavish conformity to rules. While Kant argues that it is *always* wrong to lie, for instance (even to save the life of another) he believes that a right action is one done out of "good will" or a respect for the moral law. The good person does right *because* it is right, and not for any other reason.

Kant's emphasis on respect for persons and the universality of moral rules has greatly influenced the development of moral philosophy. For instance, Harvard philosopher John Rawls' account of the just society,[1] emphasizes respect, impartiality, rationality and equality. Rawls asks us to imagine a presocial "original position" in which everyone tries to agree on rights and duties in complete ignorance of their social and economic position in the society they will set up. Indeed, this "veil of ignorance" hides from the social contractors even their good or bad luck in the distribution of talents and abilities. Rawls argues that rational contractors would certainly agree on basic principles of justice; equal basic liberty for all, and distribution of inequalities to the greatest benefit of the least advantaged.

Philosophers often use examples to bring out the distinction between different theories. Here we use the question of legal punishment. The deontological position is represented by *retributivism,* or the view that the criminal deserves to suffer solely because he or she has broken the law. By breaking the law, the argument runs, the criminal has acted unjustly, has benefited at the expense of others and failed to respect his or her fellow citizens. *Therefore* the criminal ought to be punished. In contrast, consequentialists see punishment in terms of the good it will achieve — *deterrence, protection of society*, or the offender's *reform* and *rehabilitation*. The consequentialist argues that punishment is justifiable only if it has good effects. The retributivist may welcome these results, of course, but only as side benefits. The debate on capital punishment clearly illustrates the two theories since the merits of the death penalty may be discussed in terms of the principle of "a life for a life," or in terms of the alleged deterrent effect capital punishment may have on potential murderers.

These issues often arise in discussions about environmental quality. In a case of severe environmental damage such as Love Canal or the Kepone contamination, (page 63) retributivists emphasize the injustice of passing business costs on to a helpless public via improper disposal of hazardous wastes, and demand harsh penalties in order that chemical companies pay for their alleged misdeeds. Utilitarians, concerned with public welfare, concentrate rather on compensating victims, and deterring future would-be polluters.

acts and rules

Ethical theories may further be divided into *act* and *rule* theories. An act theory requires the agent to evaluate each action individually. Thus an *act utilitarian* considers the consequences for human happiness at each occasion of choice; and an *act deontologist* tries to discover the intrinsic rightness or wrongness of each possible action. Some act deontologist theories stress the uniqueness of each occasion of choice, emphasizing the importance of the individual's life situation and relationships, the historical timing, and so on. They therefore tend towards positions of extreme subjectivism, arguing that in each situation only the individual involved can make the right decision.

Rule theories evaluate types or classes of acts rather than acts themselves. *Rule utilitarians* apply utilitarianism via a system of rules which, they believe, experience has shown to promote human happiness. Such an approach is held to simplify decision making, because obviously we cannot calculate all the possible effects of each proposed course of action. It is also argued that the existence of rules has a positive value in providing order and stability. *Rule deontologists* likewise regard *types* of actions as right or wrong. Instead of considering anew the morality of each act, the rule theorist can appeal to a set of principles, often alleged to be absolute, to guide his or her action. Thus it is *always* one's duty to tell the truth; one does not need to decide whether to lie in a particular situation. Most professional codes of ethics such as the *ASCE Code of Ethics* (Appendix A.1) include a substantial number of rules of this sort.

utilitarianism and justice

What are some of the advantages and disadvantages of these different types of theories? *Consequentialists* argue that it is always reasonable to ask why an act should be done, or avoided; that the question "What good is it?" is always, in principle, worth asking. They accuse deontologists of divorcing ethics from human welfare. They argue that the point of ethics is to benefit humans and therefore a system of ethics which requires us to perform acts without reference to or regardless of their consequences is irrational. The Scottish philosopher David Hume (1711-1776), for example attacks what he calls the "monkish virtues" of chastity, self-denial, humility and so on as causing misery without promoting anyone's well being. Jeremy Bentham criticized the notion of natural rights as "nonsense on stilts," arguing that claims of "rights" should always give way to utilitarian considerations.

Consequentialists recognize that *something* must be seen as worthwhile in itself, of course; otherwise there would be no way of deciding whether the consequences of an act are good or bad. Utilitarians believe that human happiness is good, typically basing this claim on the fact that each person values happiness above everything. In a wider sense, Aristotle (384-322 B.C.) argued that happiness is the only thing which "is always chosen for its own sake and never for the sake of something else." Thus exercise, disagreeable though it may be, is valued as a means to health, and health is good because healthy people are happier than unhealthy people.

Utilitarians claim that their theory has an empirical, practical foundation, because it identifies ultimate value with people's actual values. They argue that claims of the absolute rightness or wrongness of, say, telling the truth or murder have no basis except the alleged intuitions of the deontologist. If challenged, the latter can do little except repeat his or her certainty. In fact, as we see later, Kantians claim that ethics has a rational basis.

Nineteenth century utilitarians noted that their opponents often made appeals to some supposed ethical authority such as the Church or tradition. Such appeals are, of course, invalid. To say "You ought to do it because the Church (or tradition) says so" is to invite the question "Why should I obey the Church (or tradition)?" That some people believe they should live by the teachings of a religion is not in the least binding on non-believers. Appeals to tradition are really only appeals to convention. We try to show in Chapter 1.1 that to be ethical is not merely to follow conventions. Kant and his followers also recognize that ethics cannot be reduced to or identified with the commands of any alleged moral authority. In insisting on what he called "the autonomy of ethics" Kant is concerned to argue that the Christian should not take the view that what is right is so because God commands it. Rather, explains Kant, God commands it because (He recognizes that) it is right.

Kantians put a high value on justice and commonly criticize consequentialist theories as permitting, or even requiring injustices. Kantians argue that the only way to maximize happiness is to sacrifice the interests of some for the good of all, and that this is not acceptable. Returning to the example of punishment, retributivists claim that utilitarians would be committed to "punishing" the innocent if the result were to promote the general welfare. It has been argued that a utilitarian would favor deliberately faking evidence against a prominent American nuclear scientist, accusing him of passing secrets to the Russians, resulting in a public trial and heavy sentence, if the result would be to deter potential traitors.[2] The Australian philosopher H.J. McCloskey[3] suggests that a utilitarian sheriff in a pre-integration southern town might feel forced to arrest and unjustly condemn an innocent black for the unsolved rape of a white woman if that seemed to be the only way to avoid widespread violence by an unruly mob of white racists. By a similar process, a utilitarian regulatory agency might seek to deter potential polluters by faking evidence of breaking clean air or water regulations against an innocent firm, if the result were to reduce environmental damage by less conscientious corporations.

Does utilitarianism sometimes require one to act unjustly? A utilitarian may reply that this criticism is too shallow, for we must look beyond the immediate advantages of acting unjustly. Consider, for example, the effect on public confidence in the criminal justice system should the fraud leak out. Consider, also, the possible effects on the characters of the officials involved, who, having lapsed once, may be tempted to act corruptly in future, less urgent cases. On balance, they argue, a utilitarian *would not* favor injustice.[4]

There is, of course, the reply that whatever happens, the innocent do not deserve punishment. It can be forcefully argued that there are some things which ought not be done even if they *would* produce great benefits.

The issues here are of general relevance to environmental and other public policy decision making. A person who believed, say, that factories should not be permitted to release *any* suspected carcinogenic substances into the environment, regardless of how, where, and in what quantities, would not be prepared to trade off the advantages of cheaper production against environmental impurity.

the value of rules

Rule utilitarians can, and do, defend themselves against the charge of sacrificing principles to expediency, by claiming that rule-following itself has a positive utilitarian value. John Rawls and philosopher/lawyer Richard Wasserstrom[5] have argued that the existence of a clear, firm system of rules — such as the criminal law, or a generally accepted code of ethics — provides everyone with a certain security. We can predict the behavior of our fellow citizens, and of institutions, because we can expect them to follow rules; and regardless of the content of the rules, it is valuable to know what they are and that they will be followed. A system of rules permits and promotes impartiality and fairness, moreover, since it tends to bureaucratize justice and ethics alike. Of course, the rules should be designed so that, overall, observance of them promotes greater utility than the observance of any other rule; but even in a case where breaking the rule would lead to more desirable consequences than following it, one should obey the rule. Again, in the environmental area, rule utilitarians would not grant equitable or ad hoc exceptions to regulatory standards merely because human well-being would be advanced thereby; for to do so would be to forgo the advantages of a system of rules.

It should be emphasized that rule utilitarians usually tie the obligation to follow rules to the system's being essentially beneficent, as we show in Chapter 1.1. What if the rules are generally bad ones, and adherence to them would inhibit rather than promote happiness? Rawls believes that in general one ought to follow rules, while working through legitimate channels to get unjust rules changed; but he also accepts that there may be a right, or even a duty, of civil disobedience in the case of an extremely unjust law in a basically just system.[6]

A rule utilitarian engineer, then, might accept that rules of confidentiality should be followed, even in a case where the general welfare might be advanced by breaking the rule (e.g. disclosing the formula for some widely used product, thus reducing the price of the product and benefiting the poor). Perhaps the engineer, feeling the rule to be bad overall, should work to have it changed. But an extreme case, such as an irresponsible company intending to market some improperly tested or badly designed product, might justify "whistle blowing" — breaking the confidentiality rule. Rawls argues that the civil disobedient ought to accept the consequences, including arrest and penalty. Willingness to accept arrest and penalties shows that one is sincere and is acting on principle rather than for personal gain, thus distinguishing the civil disobedient from the criminal. Likewise, the engineer who publicly announces the company's plans (and thereby becomes known as a troublemaker) is seen to be acting on principle, as compared with the engineer who anonymously sells the story to a rival company or magazine for personal profit (and does not thereby jeopardize his or her professional career).

universalizability

Kant and some of his followers have tried to show that to be ethical is *necessarily* to follow a rule. In so doing they have tried to show that ethical principles are not merely based on appeals to intuition but have a rational basis.

Kant holds that one way to bring out the wrongfulness of a type of behavior is to show that it could not be adopted by a rational being. For instance, he tries to show that it is irrational to choose to cheat in business, or to fail to help others in distress. The cheat adopts as a self-interested maxim: Always cheat when I can get away with it. But, Kant maintains, a test of the rationality of a moral principle is: Can it be *universalized*? One cannot, as a rational being, adopt a rule for oneself unless one is prepared to accept it as applicable to and by *all* persons. But since no rational being wants to be cheated (which would have to be the case, were the maxim to be universalized), the policy of cheating others turns out to be irrational. The same can be said of a policy of ignoring others who are in distress. What first appears to be in one's self-interest turns out to be opposed to it.

Kant's theory of the universalizability of ethical judgments is but one version of what he calls the *categorical imperative*. He also presents it in what he asserts to be merely different versions with essentially the same meaning. One of these is the *Golden Rule*; do as you would be done by. Another has come to be known as the principle of *respect for persons*. The core of this principle is that one must always treat other persons as *ends*, and never as mere *means*. Once again, cheating in business is immoral since one thereby treats one's customer as mere means to one's own enrichment. Because one must treat humanity in one's *own* person as an end, with respect, it follows that there are duties to oneself. For example, suicide is immoral by this formulation too, since to kill oneself is to treat one's life as a mere means — such as to secure release from pain. Likewise, it is immoral to refuse to develop one's talents out of laziness.

R.M. Hare, a contemporary Oxford philosopher in the Kantian tradition, develops the universalizability principle in *The Language of Morals*.[7] Hare argues that the distinctive feature of a moral principle (as opposed to a mere prejudice or a resolution to indulge oneself) is its general nature. To call a thing good is to commend it to others. To assert that an action or principle is right is to commit oneself to doing it, and to prescribe it as a duty to others. It would be quite irrational to assert, for example, that one had a right to accept bribes, and yet deny that anyone else had that right. In order to make use of ethical terms such as ''right'' one must recognize that one commits oneself to the assertion of general moral principles. To claim that I uniquely have the right to accept bribes is to place myself in a specially favored category — which I cannot justify without giving special reasons. By giving reasons I am (whether I like it or not) asserting that *anyone in my position* has the right too. To claim the right to accept bribes is to assert bribery as a norm, which few would welcome.

Kant assumes that everyone has a rational perception of his or her own interests, so that the mere recognition that the adoption of a principle by all would work against one's own interests would suffice to cause one to abandon the principle. On balance, it might be said, even purely self-interested people will prefer to limit their pursuit of self-interest provided that everyone does. Hare[8] discusses the case of what he calls the *fanatic*, who insists on adhering to a principle even where it would work *against* his or her interests; an example would be a Nazi who asserts that should he turn out to be Jewish then it would be right to send him to the gas chamber. Faced with that possibility, the Nazi must either abandon the principle or accept that his own death would be a good thing. A valuable feature of this sort of example is that it should make us very suspicious of all cases of sweeping and absolute principles, especially where they appear unjust. Of course, once the Nazi exhibits his fanaticism we cannot reason any further, but we have at least exposed the implications of his belief.

The "Golden Rule" implications of Kantian theory are brought out in Hare's discussion of abortion.[9] He argues that we would not have wished our parents to have aborted us (had we been in a position to wish) and are indeed glad that they did not abort us; therefore we ought not to support abortion on demand. For if it is right to abort a fetus merely because it is an inconvenience, then it would have been all right, for anyone, ourselves included, to have been aborted. (Not everyone will be persuaded by this argument, of course. To wish that one had never been born may be intelligible or even rational if one's life is totally miserable.)

The concept of respect for persons can be similarly explained. Nobody wants to be treated as a mere means to the ends of others; we all wish to be respected as valuable autonomous beings with our own goals, which we are entitled to pursue. But if this is how we would have others treat us, we ought to accord them the same respect. If we claim it as our right, we are then logically bound to respect the rights of others also, unless we can show that we have special qualities entitling us to respect which others lack.

is ethics no more than rules?

Many philosophers have been dissatisfied with an account of ethics in terms of rules. *Situation ethics*[10] stems from a rejection of the rather rigid ethics as rule following taught by many conventional Christian institutions. This theory puts ethical responsibility on the agent; right action is what one is motivated to do out of love for others. The proponents of situation ethics emphasize what they take to be central to the true Christian ethic; a free, loving response to the needs of others.

Existentialist criticism of the Kantian tradition focuses on the uniqueness of each choice, and particularly on the responsibility of each person to make his or her own moral choices. The value of choice, for the existentialist, is its courageous assertion of humanity and autonomy in a meaningless universe. The universe is meaningless, according to the French philosopher Jean-Paul Sartre

(1905-1980) because there is no God, no grand plan or design, no system of final rewards and punishments; each therefore has the awesome responsibility of choosing for himself or herself, and thereby defining themselves as persons. To take refuge in conventional morality or other rules, to refuse to make one's own choices autonomously, is to fail as a human being.[11]

Existentialists are fond of illustrating ethics by reference to bizarre situations, where no conventional solution presents itself. The French/Algerian writer and resistance hero Albert Camus (1913-1960) presents such situations in his plays and novels. In Camus' *The Plague*[12] the main character, a doctor, decides to stay in a city swept by the plague in order to carry out the inevitably hopeless task of taking care of the victims. His decision is neither reasonable nor practical, but it is *right* — for him.

Existentialists are also fascinated by situations in which conventional morality is no use as a guide to action because it offers several conflicting solutions, and by dilemmas or situations in which the individual is forced to choose between equally undesirable alternatives. A famous example of Sartre's is the dilemma of a young man in German-occupied France, faced with the contradictory obligations of taking care of his elderly mother and of joining the underground resistance. Each of these actions, considered by itself, is a duty; each carries an unacceptably high price. In these situations, individuals are thrown back on their own moral resources.

To act ethically, for an existentialist, is to accept that one *must* choose for oneself; that the act of choice is the supreme creative and life-affirming act; and that this is one's greatest responsibility, because to choose is to commit oneself to living by the consequences of one's choice. The choice between staying in the plague-ridden city and leaving for safety is a final commitment — one cannot renege.[13]

Existentialists reject the Kantian tradition of universalizability; as the British philosopher and writer Alasdair MacIntyre points out,[14] that tradition emphasizes the *similarities* between types of situations and kinds of people. For the existentialist, the only person relevantly similar to the agent *is* the agent; the characterization of moral problems includes the person making the choice, so that analogies, rules and principles serve only to divert one from one's responsibilities.

Existentialism and situation ethics contain valuable insights; in particular they help us to avoid oversimplification. They may help us to resist the temptation to overemphasize some element of human nature such as rationality (Kant) or the desire for happiness (the utilitarians). They are also a useful corrective to the "plug and chug" view that ethics consists of fitting a situation into a rule, upon which one then acts, like an unimaginative physician diagnozing a condition and prescribing the usual pills. In a wider context, technology optimists and environmental crisis pessimists are united in the view that conventional solutions to environmental problems are no longer adequate, and it may be that our traditional economic and political framework is inappropriate in an age of overpopulation and nuclear weapons. In the section on environmental ethics (Chapter 1.2), we suggest that traditional ethics, likewise, may be inadequate to deal with environmental issues in a changing world.

Critics of existentialist and situation ethics claim that the theories overemphasize the uniqueness of each situation, possibly to the exclusion of cooperation and joint action. Perhaps situation ethics is most appropriate for ethical questions of personal relationships and as an expression of concern for outcasts and misfits for whom no conventional provision can be made. Existentialism may offer guidance for the person seeking a personal goal or ideal, or for situations where there is no hope of agreed or institutionalized solutions to problems.

conclusions

If one were to try to construct an ethic for the professional engineer, building on the Western tradition of ethics, it might look something like this: To be ethical is to take the universal point of view, to have respect for all persons, to take account of the effects of our actions on their welfare and other interests, and when in doubt apply the Golden Rule. At the same time, the ethical person recognizes that he or she has a considerable personal responsibility to choose rightly and not merely to accept conventional practice.

The duty of a professional engineer thus goes beyond pursuing his or her own career and giving satisfaction to employer or client. The ethical engineer, in respecting the rights of others and the interest of the community (ultimately all humanity), will not advance his or her career at the expense of client, employer or community. Nor, we suggest, should the interests of client or employer be put above those of the society at large which, after all, makes possible the practice of engineering and for whose benefit it ought to be practiced. And finally, the ethical engineer must take personal moral responsibility for his or her actions. To live ethically is thus to have self-esteem; to feel proud of oneself for doing what is *right*.

references

1. Rawls, J. *A Theory of Justice*, The Belknapp Press of Harvard University, Cambridge MA, 1971.
2. Armstrong, K. "The Retributivist Hits Back" *Mind*, 70, 471-490, 1961.
3. McClosky, H.J. "A Non-Utilitarian Approach to Punishment" *Inquiry*, 8, 249-263, 1965.
4. Sprigge, T.L.S. "A Utilitarian Reply to Dr. McCloskey" *Inquiry*, 8, 264-291, 1965.
5. Wasserstrom, R.A. *The Judicial Decision*, Stanford University Press, Stanford CA, 1961.
6. Rawls, J. "The Justification of Moral Disobedience" in *Today's Moral Problems*, R.A. Wasserstrom (Ed.), McMillan, New York, 1975.
7. Hare, R.M. *The Language of Morals*, Oxford University Press, London, 1952.
8. Hare, R.M. *Freedom and Reason*, Oxford University Press, London, 1963.
9. Hare, R.M. "Abortion and the Golden Rule," *Philosophy and Public Affairs*, 4, 401-222, 1975.
10. Fletcher, J. *Situation Ethics: The New Morality*, SCM Press

Limited, London, 1969.

11. Sartre, J-P., *Existentialism and Humanism*, Methuen, London, 1963.

12. Camus, A. *The Plague*, Penguin Books, Harmondworth, England, 1963.

13. Sartre, J-P., as quoted in McIntyre, A. "What Philosophy is Not," *Philosophy*, 32, 332-335, 1957.

14. McIntyre, A. "What Morality is Not," *Philosophy*, 32, 332-355, 1957.

Index

A

acid rain 18
act deontologists 139
Adams, H. 97
Adams, Pres. J. 60
Amish 24
Allied Chemical 63-71
Aristotle 19, 140

B

Bachelelard, G. 93
Bacon, F. 22
Baeder, D. 73-85
Bentham, J. 137, 140
Boulle, P. 97
Bunker Hill ID 87-89

C

Camus, A. 147
Cape Fear River 101
Carnahan, C. 75, 77
Carter, Pres. J. 106
categorical imperative 144
CBS News, 60 Minutes 73-85
Chou, Dr. 70
Code of Ethics 33, 117, 132-135, 140
cognitive dissonance 19
consequentialist ethics 137, 140
Cooley, Rep. H. 102
Corps of Engineers 34, 65, 66, 92, 101-109, 111

D

Darwin, C. 123
Den Doolaard, A. 94
Descartes, R. 22
deontological ethics 137-138
Deukmejian, G. 83
Douglass, Justice W. 101
Dubois, W. 128

E

Edson, R. 73-77
Endy, E. 111-121
environmental impact statements 35-37, 102
Environmental Protection Agency (EPA) 66, 112
ethical egoism 137
Everglades 34-35
existentialism 146-148

F

Fayetteville NC 101
Florman, S. 91-99, 101, 116
Frank, W. 96

G

Galilei, G. 22
garden ethic 21
Garner, J. 102
golden rule 31, 141, 144, 146, 148
Goldfarb, W. 63-71
Golding, W. 97
Gore, Rep. A. 82-83
Guthrie, D. 79-80

H

Halley, J. 89
Hare, R. 145
Hardin, G. 55-61, 123
Harris, R. 77-78
Haw River 101-102
Hobbes, T. 22
Holmes, Justice W. 127
Hooker Chemical Co. 64, 67, 73-85
Hoover Commission 106
Hopewell VA 63-71
Hume, D. 140
Hundtofte, V. 65-68, 70
Hustable, A. 95
Hydrocomp 104, 107

J

Jackson, Dr. R. 70
Johnson, Pres. L. 106
Johnstone, Col. H. 107-109
Jones, C. 68
Jordan, Sen. B. Everett 102
B. Everett Jordan Dam 35, 101-109

K

Kant, I. 138, 141, 144, 146-148
Kentucky 46-47
kepone 63-71, 139
Kohlberg, L. 113-120
Kriedeman, K. 88-89

L

land ethic 24, 43-53
Lathrop CA 74-85

lead 87-89
Leopold, A. 24, 30, 43-53, 123
Life Science Products Co. 67-71
Lloyd, W. 56
Love Canal 73, 139
Lorax, The 34

M

MacIntyre, A. 147
Maine, H. 124
Maori 38
Marshall, Justice 125
McCloskey, H. 141
McCuen, R. 115
mechanistic view of nature 22
Michener, J. 94
Moore, W. 65-68, 70
Montefiore, H. 21
moral community 17
moral development 112-121
Mosaic Decalogue 44
Moses, H. 77
Murie, M. 26

N

National Environmental Policy Act 35, 102
National Parks 58
National Pollutant Discharge Elimination System 66
Nease Chemical Co. 64, 67
New Hope Creek 102
New Hope Dam 35, 102
New Mexico 47
New Zealand 38
Niagara Falls NY 73
Nixon, Pres. R. 35

O

Occidental Chemical Co. 73-85
Okeechobee Lake 34
Osborne, A. 76

P

Piaget, J. 113-120
Public Works Committee 106
Pueblo 47

R

Rawls, J. 138, 142-143
Refuse Act Permit Program 65
Regan, T. 29
Roosevelt, Pres. F. 106
Rousseau, J.-J. 95
rule utilitarian 139
rule deontologist 140
Rutgers University 64

S

Sartre, J.-P. 146
Scott, Sen. 102
Seuss, Dr. 34
sexism 20
Singer, P. 28
situation ethics 146
slavery 19-20, 43
Smeaton, J. 94
Smith, A. 56
soil conservation 48

speciesism 22-24, 30
standing 33, 123-129
Stevenson, R. L. 93
Stone, C. 123-129
Story, Justice 125

T

Tate, C. 87-89
Ten Commandments 138
teleological ethics 137

U

U.S. Army Corps of Engineers 34, 65, 66, 92, 101-109, 111
universalizability 30-31, 144-146
utilitarianism 28, 137-150

V

Valery, P. 92
Velsicol 67
Vesilind, P. A. 101-109, 111-121
Voting Rights Act 124

W

Wallace, M. 73-85
Wasserstrom, R. 142
Weiser, E. 102
whistle blowing 134
White, L. 21
Whitehead 56
Wilkenfeld, J. 79-80
Wisconsin 48, 128

Alastair S. Gunn is a Lecturer at the University of Waikato in Hamilton, New Zealand, having emigrated from his native Great Britain in 1969. He earned his BA in philosophy and completed Graduate Study at the University of Sussex. With his main teaching interests in applied ethics, he teaches or has taught business and professional ethics, engineering and environmental ethics, medical and police ethics, women's studies and philosophy.

The author has been visiting scholar or guest lecturer at the Duke University Environmental Center, North Carolina State University, and the University of Auckland Engineering School.

Active in Australasian associations, the author has served as Editor of *Waikato Environment* (journal) and as President of both the University of Waikato Lecturers' Association and the New Zealand Division of the Australasian Association of Philosophy.

132

P. Aarne Vesilind, P.E. is Professor and Chairman of the Department of Civil and Environmental Engineering at Duke University, Durham, North Carolina, and has been Director of the Duke Environmental Engineering Center since 1974. Born in Estonia, he holds BS and MS degrees in civil engineering from Lehigh University, and an MS and PhD in sanitary engineering from the University of North Carolina. Dr. Vesilind was a Fulbright-Hayes Senior Lecturer at the University of Waikato in Hamilton, New Zealand.

Professor Vesilind has done design and consulting work for various industrial and consulting firms, and for government agencies.

With this book, he is now author or co-author of 12 books, the eleventh being *Sludge Management and Disposal for the Practicing Engineer* (with Gerald C. Hartman and Elizabeth T. Skene), published early in 1986. Professor Vesilind is author of more than 100 research papers and articles, and serves on the Editorial Board of Lewis Publishers, Inc.